AF544948

INTEGRATED PEST MANAGEMENT (IPM) - ENVIRONMENTALLY SOUND PEST MANAGEMENT

Edited by **Harsimran Kaur Gill**
and **Gaurav Goyal**

Integrated Pest Management (IPM): Environmentally Sound Pest Management

http://dx.doi.org/10.5772/61693
Edited by Harsimran Kaur Gill and Gaurav Goyal

Contributors

Ugur Gozel, Cigdem Gozel, Maria Pojar-Fenesan, Ana Balea, Thanh-Danh Nguyen, Chi-Hien Dang, Cong-Hao Nguyen, Chan Im, Vasile Stoleru, Vicenzo Michele Sellitto, Yelitza Coromoto Colmenárez, Carlos Vásquez, Natalia Corniani, Javier Franco, Tomislav Curkovic, Vladimir Puza, Zdenek Mracek, Jiří Nermuť

CBS, Edition **2017**

Published by InTech

Janeza Trdine 9, 51000 Rijeka, Croatia

Publishing Process Manager Andrea Koric

Technical Editor SPi Global

Cover Designer InTech Design Team

Additional hard copies can be obtained from orders@intechopen.com

Integrated Pest Management (IPM): Environmentally Sound Pest Management, Edited by Harsimran Kaur Gill and Gaurav Goyal
p. cm.
Print ISBN 978-953-51-2612-6
Online ISBN 978-953-51-2613-3

INTECH
open science | open minds

Contents

Preface

The idea of pest control is rarely discussed without referring to the concept of integrated pest management, or most commonly used as IPM. IPM is a holistic approach for pest control that seeks to optimize the use of combinations of different methods or options to manage a whole pest spectrum in particular cropping system while minimizing risks to people and the environment. Several techniques of pest management have been used in the past and are currently being used by farmers, researchers, and others, but it is even more important in the current global pest scenario to utilize the pest management strategies in an effective and holistic approach for them to be available in the future. Very often a pest management approach gets discovered and then over-utilized to such an extent that it comes at the verge of extinction due to either development of resistance against it or environment impact of this. Moreover, individual pest management techniques aim toward bringing the pest population down to a level, which is either not cost-effective or not sustainable in the long run, and the pest population tends to bounce back once the control measure is taken away.

IPM approaches the pest management in such a way that the pest management is achieved by utilizing many pest management strategies in a way that the control achieved is cost-effective and sustainable over generations. Since the idea is to bring the pest population below threshold level, IPM is more sustainable. The pest population remains suppressed for long time and therefore prevents the need for frequent pest management and therefore reducing cost. Many pest management professionals rely on chemicals to control the pests. IPM therefore aids in protecting environment by preventing the use of many chemicals and instead relying on the combination of pest management approaches. Some of the pest management techniques used in IPM are biological control, cultural control, mechanical control, physical control, chemical control, etc. IPM is used in agriculture, horticulture, structural pest management, turf pest management, ornamental pest management, and human habitations. Insect pest management is a subsystem of IPM, and these two terms are used as synonymous most of the time.

The book *Integrated Pest Management (IPM) - Environmentally Sound Pest Management* is intended to provide an overview of eco-friendly options for pest management in agricultural cropping systems. Chemicals have been long used worldwide in the past for management of agricultural pests. Due to their potential negative effects on human health; environment including soil, water, and air; and others, chemicals have to be used very judiciously. Private companies are developing the pest management chemicals themselves or through third-party contracts including public universities that produce huge amount of data on chemical efficacy and safety before chemical registration and the use of chemicals in agriculture. Comparatively not much attention is given to other control methods, which are either because those control methods are not very effective compared to chemicals or not much research has been done to improve that method. The book focuses on some of those pest management methods that have been employed worldwide highlighting the major problem and issues and possible attempts to identify promising lines and directions for future research and implementation. Many researchers have contributed to the publication of this book. We aimed to compile information from a wide diversity of sources into a single volume in forming this book. We begin with historical review of IPM concepts, strategies, and some experiences in applications of IPMS in Latin America. The rest of the six chapters offer information on pest management approaches alternative to chemicals. The chapters include pest control in organic agricultural system through preventive and curative measures; the use of entomopathogenic nematodes in pest management; advances in production, storage, application techniques, genetic improvement, and safety of entomopathogenic and molluscoparasitic nematodes, which are important parasites of many insect and mollusks, respectively; review of performance of popular insect pheromones used in Vietnam; semiochemicals use in IPM environmentally compatible strategies to reduce pest population under economic threshold levels; and management of agriculture pests using detergents and soaps as parts of IPM scheme.

The inclusion of different methods for pest management globally will make this book of significance to researchers, scientists, graduate students, growers, policy makers, and other professionals who can make use of compiled information from this book. Environment safety is one of the top concerns these days with growers either looking for or forced by policy makers toward more environment-friendly options than ever before. This book is not intended to provide all the alternative pest management methods but to provide many of the common ones evaluated by researchers and with feasibility over grower's farm. We hope that this book will continue to meet the expectations and needs of anyone interested in the topic to learn more and understand different IPM options.

Harsimran Kaur Gill, PhD,
Cornell University,
Ithaca, NY, USA

Gaurav Goyal, PhD,
Monsanto Company,
St. Louis, MO, USA

Chapter 1

Implementation and Adoption of Integrated Pest Management Approaches in Latin America: Challenges and Potential

Yelitza Colmenárez, Carlos Vásquez,
Natália Corniani and Javier Franco

Additional information is available at the end of the chapter

http://dx.doi.org/10.5772/64098

Abstract

Latin American countries present diverse agricultural systems, ranging from the subsistence agriculture in common property lands to large highly mechanized estates that produce crops for export. Despite this diversity, the adoption of integrated pest management (IPM) is commonly based on reducing the negative effect of pesticides on consumer health and on the environment. In most of Latin American countries, the agricultural sector is characterized by poor infrastructure in research and extension systems, a public sector with limited human resources that limits the dissemination of information and provides inappropriate credit and subsidy schemes, all of these have influenced negatively on the possibility of the success of IPM programs. Thus, some innovative alternatives have emerged from concerning public and private initiatives. In this regard, the Plantwise approach, as a framework for action, is to strengthen the capacity of agricultural institutions and organizations to establish more effective and sustainable national plant health systems. Plantwise is an innovative global program led by the Centre for Agriculture and Biosciences International (CABI), which aims to contribute to increased food security, alleviated poverty, and improved livelihoods by enabling male and female farmers around the world to lose less, produce more, and improve the quality of their crops. Strengthening plant health systems removes barriers to make accessible to farmers sustainable approaches for pest control. In this chapter, we include some historical review of IPM concepts, strategies, and some experiences in application of IPM in Latin America. Also we discuss the potential and challenges for implementation and adoption of IPM practices and the ways how Plantwise has engaged with the key partners in the different countries where the program is being implemented, promoting the implementation of IPM approaches in order to improve agriculture systems, mainly those from subsistence agriculture, in Latin America.

Keywords: integrated pest management, Latin America, Plantwise, plant health systems

---

1. Introduction

One of the main challenges of the agriculture is to provide increasing supplies of food for a growing population with the increase in efficiency in the use of inputs and reduction of the environmental impacts from production [1], where both the ecological and economic dimensions are considered [2]. In this context, Yudulmen et al. [3] stated that the efficient management of insect pests should have a high priority given that insects still take about 15% of potential global crop yields [3]. However, the use of pesticides as one of the major control strategy adds economic and environmental costs to the food production equation [4].

The Integrated Control Concept (ICC) created by Stern [5] gave rise to the idea of the integrated pest management (IPM) and it has been a scientifically accepted "paradigm" for pest management worldwide for more than 50 years. In the context of the ICC, a fundamental element is to understand that any control system imposed on a given pest in a given crop has consequences for the management of other pests and crops in the ecosystem [6]. Thus, the IPM is a multitactic nature approach, including aspects related to host plant (such as plant nutrition, plant physiology, and plant resistance) and the economic aspects.

Considering the pyramidal conception of an IPM program designed for whitefly management (**Figure 1**), it is possible to generalize that model to other pests. In this regard, we could state that avoidance constitutes the basis of a pest management program, although some might reside on more than one level. For example, when facing a pest outbreak, decisions could be made based upon the upper two levels of the pyramid.

Later, Kogan [8] defined IPM as a decision support system for the selection and use of pest control tactics, singly or harmoniously coordinated into a management strategy, based on cost/

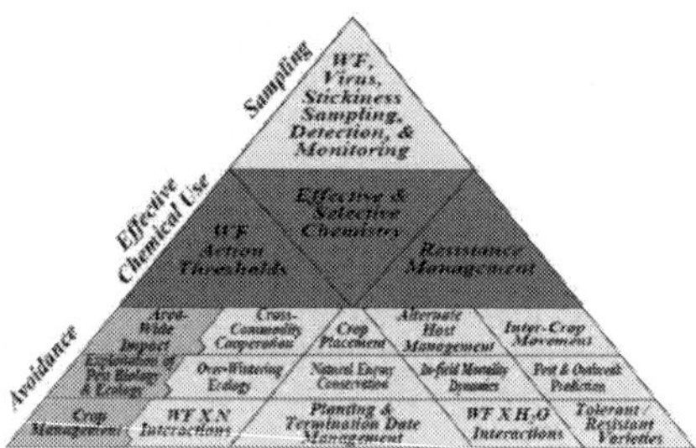

Figure 1. Conceptual diagram of whitefly IPM, depicting three keys to whitefly management (left): sampling, effective chemical use, and avoidance. Avoidance is subdivided among three interrelated areas: area-wide impact, exploitation of pest biology and ecology, and crop management (from: [7]).

benefit analyses that take into account the interests of and impacts on producers, society, and the environment. According to Rodríguez and Niemeyer [9], this definition inherently considers the existence of ecological and economic thresholds, the need to adopt the socioecosystem as a management unit, the existence of a broad number of IPM tools including the rational use of chemical pesticides, and the requirement for interdisciplinary systems approach, particularly since certain control measures may produce unexpected and undesirable effects. As complement to the classical definition, the United States Department of Agriculture has defined the IPM as a long-standing, science-based, decision-making process that identifies and reduces risks from pests and pest management–related strategies [10].

Additionally, Naranjo and Ellsworth [6] discussed the evolution of IPM concepts built on the original four components: thresholds for determining the need for control, sampling to determine critical densities, understanding and conserving the biological control capacity in the system, and the use of selective insecticides or selective application methods, when needed, to augment biological control.

2. Plant protection techniques used in IPM

IPM relies mainly on natural mortality factors such as natural enemies and weather seeking out tactics that disrupt these factors as little as possible [11]. In a broader sense, it includes all plant protection measures that help to prevent or manage pests, whether through general crop management practices such as rotation, or of cultural, physical, biological, or chemical nature. When pesticides are applied, two crucial items to be considered are determining when pesticides actually need to be used and the choice of chemicals should be made with consideration of compatibility with nonchemical methods (e.g., natural predators), pest population level, and resistance management, products' profiles. In an IPM context, these decisions are heavily based on an important step such as the biological monitoring (also referred as 'scouting'), which consists of sampling procedures designed to estimate the stages and population densities of both pests and beneficial organisms [12]. Unfortunately, biological monitoring is a very knowledge-intensive procedure and requires highly trained individuals to obtain reliable data and consequently ensure the success of the program. On the other hand, since both pest populations and the growth and development of crop plants are governed by environmental parameters, monitoring environmental conditions should be another core component of IPM [12].

For all that, crop production is dynamic; the decisions on pest management measures should be taken at farm level based on a wide variety of instruments, such as qualified advisers' recommendations, alert services and infestation forecast, research results, experience, and threshold values. However, the actual techniques to be included in an IPM approach on-farm will vary not only between crops but also within the same crop grown in different geographical locations, or between years, depending on pest pressure, weather patterns, crop rotation, and other factors, as well as availability of tools and resources [13]. All these should consider economic aspects, trying to allocate scarce resources (capital or labour) [14].

3. Biological control and IPM

Biological control has been a valuable tactic in pest management programs around the world for many years, but has undergone a resurgence in recent decades that parallels the development of IPM as an accepted practice for pest management [15]. Since natural enemies are often key factors in the dynamics of pests, biological control should be the cornerstone of IPM practices [16]. However, when implementing an integrated pest management programs, special care should be taken in what specific tactics could be used since they do not act independently of one another. This is especially true for biological control since the agents of insect biological control are susceptible to environmental factors, such as pesticides, cultural control, mechanical and physical control, and transgenic crops [15].

However, both biological control and IPM faced some obstacles originating from the lack of biological data and the lack of knowledge to develop economically, environmentally, and socially sound crops and animal production systems [17].

Insect or mite species	**Developmental stage attacked**	***Rhynchophorus* species**	**Location**
Insects			
Anisolabis maritime (Dermaptera: Anisolabididae)	Eggs, larvae and pupae	*R. ferrugineus*	Saudi Arabia
Chelisoches morio (Dermaptera: Chelisochidae)	Eggs and larvae	*R. ferrugineus*	India
Euborellia annulipes (Dermaptera: Anisolabididae)	Eggs	*R. ferrugineus*	Italy
Platymeris laevicollis (Hemiptera: Reduviidae)	Unknown	*R. ferrugineus*	Sri Lanka
Xylocorus galactinus (Hemiptera: Anthocoridae)	Eggs, larvae and pupae	*R. ferrugineus*	Saudi Arabia
Xanthopygus cognatus (Coleoptera: Staphylinidae)	Eggs and larvae	*R. palmarum*	Ecuador
Sarcophaga fuscicauda (Diptera: Sarcophagidae)	Adults	*R. ferrugineus*	India
Billea rhynchoporae (Diptera: Tachinidae)	Pupae	*R. palmarum*	Brazil
B. maritima	Pupae	*R. ferrugineus*	Italy
B. menezesi	Pupae	*R. palmarum*	Brazil
Megaselia scalaris (Diptera: Phoridae)	Pupae	*R. ferrugineus*	Italy
Scolia erratica	Larvae	*R. ferrugineus*	Malaysia

Insect or mite species	Developmental stage attacked	*Rhynchophorus* species	Location
(Hymenoptera: Scolidae)			
Mites			
Aegyptus alhassa (Mesostigmata: Trachyuro podidae)	Eggs, pupae and adults	*R. ferrugineus*	Saudi Arabia
A. rynchophorus	Pupae and adults	*R. ferrugineus*	Egypt
A. zaheri	Pupae and adults	*R. ferrugineus*	Egypt
Uroobovella marginata (Mesostigmata: Urodinychidae)	Pupae and adults	*R. ferrugineus*	Egypt
Hypoaspis sardoa (Mesostigmata: Laelapidae)	All stages	*R. ferrugineus*	Egypt
Hypoaspis sp.	Adults	*R. ferrugineus*	India
Iphidosoma sp. (Mesostigmata: Parasitidae)	All stages	*R. ferrugineus*	Egypt
Parasitis zaheri (Mesostigmata: Parasitidae)	Larvae and pupae	*R. ferrugineus*	Egypt
Rhynchopolipus rhynchophori	Larvae	*R. ferrugineus*	India
(Prostigmata: Podapolipidae)	Adults	*R. palmarum*	Central and South America, Costa Rica
R. brachycephalus	Adults	*R. phoenicis*	Cameroon
R. swiftae	Adults	*R. ferrugineus*	Indonesia, Malaysia, Philippines

Table 1. List of insects and mites as natural enemies of *Rhynchophorus* spp. worldwide (from Mazza et al. [19])

The red palm weevil, *Rhynchophorus ferrugineus* (Olivier) (Coleoptera: Curculionidae), is a well-known problem for the damage it causes to coconuts (*Cocos nucifera*) grown in plantations so that much research has been conducted with a strong emphasis on the development of IPM based on pheromone traps and biological control rather than insecticides [18]. Thus, these authors stated that the prospects for the development of a biological control component for an integrated management strategy are good; however, the establishment and effectiveness of the biological control may depend on the intensity of management practices in palm (*Phoenix dactylifera*) plantations. In addition, there is also scope for the development of biopesticides to replace directly or to reduce the use of chemical pesticides. In this regard, Mazza et al. [19] have showed a list of insects and mites as natural enemies of *R. ferrugineus* worldwide (**Table 1**). As shown, most diverse insect groups belong to Diptera (4 spp.) and Dermaptera (3 spp.), while in the group of mites, Mesostigmata are the dominant species group. Regarding geographical distribution, most of the studies have been conducted in Egypt and in some countries from Asia (India, Indonesia, Malaysia, Philippines, Saudi Arabia, Sri Lanka) and most discrete number of studies in Latin America, with reports from Brazil, Costa Rica, and

Ecuador. This fact reveals the limited information about natural enemies in Latin America, thus making difficult to establish IPM programs with a predictable success opportunity. Thus, more studies concerning the biological parameters of the pests and their natural enemies are required in this geographical area.

Another successfully pest control program, known as the Moscamed Program, was developed in Mexico with participation of Mexican and Guatemalan authorities and the USDA in collaboration with the FAO and International Atomic Energy Authority (IAEA) to manage the Mediterranean fruit fly (*Ceratitis capitata*). The Moscamed program involved the application of insecticidal baits, mechanical and cultural control of hosts, restrictions on the movement of fruits and vegetables and the release of sterile males produced in the Moscamed plant at Metapa, Chiapas [20].

4. IPM in some Latin American countries: successful experiences

In South America, IPM has been successfully implemented in Argentina [lucerne (*Medicago sativa*), citrus (*Citrus* sp.), soybean (*Glycine max*)], Brazil [(citrus (*Citrus* sp.), cotton (*Gossypium* sp.), soybean (*G. max*), sugarcane (*Saccharum officinarum*), tomato (*Solanum lycopersicum*), wheat (*Triticum vulgare*) and livestock)], Chile [wheat (*Triticum vulgare*)], Colombia [cotton (*Gossypium* sp.), ornamental (*Rosa* sp.), soybean (*G. max*), sugarcane (*Saccharum officinarum*), tomato (*Solanum lycopersicum*)], Paraguay [cotton (*Gossypium* sp.), soybean (*G. max*)], Peru [cotton (*Gossypium* sp.), sugarcane (*Saccharum officinarum*)], and Venezuela [cotton (*Gossypium* sp.), sugarcane (*Saccharum officinarum*)] [20].

4.1. Argentina

Since the 1970s, Argentinian public institutions started to introduce farmers to IPM strategies by implementing a program of Extension and Technology Transfer focusing on the rational use of pesticides [21]. Although other IPM programs in soybeans, potatoes, and orchard crops have been developed, the cotton IPM program is being the oldest program. In this cotton IPM program, some strategies such as conservation of natural enemies, prevention of pesticide resistance, and cultural practices have been used.

At the beginning of the IPM program, farmers and technicians were trained for insect identification and monitoring training, however, few growers put the knowledge into practice. As a consequence of the severe economic problems caused by the lack of control of *Alabama argillacea* (leafworm) in cotton (*Gossypium* sp.), a new technology transfer program was organized to teach IPM philosophy and thus the Cotton IPM Program reappeared [21]. According to these authors, after this fact, farmers understood that adequate insecticide use at the proper timing and at the correct dose reduces costs of production and provides more efficient crop management.

4.2. Brazil

Pesticide resistance, pest resurgence, worker poisoning, and ecological imbalances became apparent after indiscriminate pesticide usage in Brazil. In this regard, research was carried out on sampling methods on pests and natural enemies, use of threshold levels, and the correct timing for insecticide application [22]. Consequently, highly successful IPM programs were developed for several crops, including sugarcane, tomato, wheat, and soybean [23].

According to Hoffmann-Campo [23], most IPM programs in Brazil are characterized as follows:

a. IPM is strongly based on using on the production and release of biological control agents with new IPM programs being developed making an emphasis on conservation and augmentative biological control, cultural practices, and host plant resistance and emphasize the reduction of broad-spectrum insecticide use.

b. Considerable improvements are expected in the methods of production and release of indigenous entomopathogens, parasitoids, and predators. Some systems are exploring classical biological control.

c. Brazilian farmers are increasingly using safer and more selective insecticides, such as the biological, the insect growth regulators (IGRs) and nicotinoids and other new products released by private companies. IPM tactics are increasingly used in Brazil since more high-quality food and fewer chemical pesticides used in food production are currently demanded by consumers and also due to the policies for registration and use of insecticides in the country have become more stringent.

d. Organic farms are a growing sector in Brazil, with an increasing demand for pest control methods that can be used on organic crops.

e. Although continuous development and improvement of IPM programs in Brazil is important, improved technology transfer and outreach to growers is fundamental. After introduction of the Genetically Modified Organisms (GMOs), research on their application to IPM programs is underway, especially their impact on natural enemies, nontarget insects, and other arthropods that feed on these crops, as well as the possibility of pest resistance.

IPM tactics must be made widely available to farmers through research institutions, official and private (farmer's cooperatives) extension services, and private companies. It is only by educating farmers on the importance and benefits of using IPM tactics for pest control that IPM programs can have a broader impact on agriculture in Latin America.

4.3. Ecuador

Information about IPM in Ecuador is still scarce. However, some attempts have been done mostly in cocoa (*Theobroma cacao*), sugar cane (*Saccharum officinarum*), and vegetable crops. In Ecuador, about 500,000 ha are planted with cocoa cultivars 'CCN-51' and 'Nacional'. Defoliating insects belonging to Saturdinae and Megalopygidae (Order: Lepidoptera) commonly infest these cultivars. When high population levels are attained in adult plantations, control by broad-spectrum insecticides application is limited since populations of pollinators can be affected. Foliage application of biological pesticide *Bacillus thuringiensis* (New BT 2X at a rate 0.5 kg ha^{-1} or New BT 8L at 1 L ha^{-1}) has showed promissory results in control of these lepidopteran pests [24].

Sugar cane: Program for the development of IPM from CINCAE (Centro de Investigación de la Caña de Azúcar del Ecuador) has proposed the following program [25]:

a. During the first phase, an evaluation and characterization of pests to determine the impact (population, damage, and grower's perception), followed by bioecological studies (life cycle, behavior, and population dynamic).

b. After that, some management components should be developed, focusing in methods of control that provoke more permanent natural mortality, being pesticides the last strategy to be considered. When pesticides are used, the minimum number of applications of selective molecules should be considered. After that, key components are integrated in a basis ecological, agronomical, and socioeconomically compatible.

Finally, pilot units are settling down in fields where these compatible components are used according to the characteristics of each agroecosystem.

4.4. Mexico

Mexico has a long history of proactive pest management, and more recently, IPM has become even more important as trade regulations that have begun to restrict the amounts of pesticide residue or insects that may be present on produce exported to the USA and Canada [26]. In order to maintain the extensive trade in fresh fruits and vegetables, these commodities must comply with strict regulations that are difficult to meet with conventional pest control methods, being IPM, in most of the cases, the only viable option for growers intending to export their products [26]. Several IPM programs have been successfully developed in Mexico.

IPM to control the tomato pinworm, *Keiferia lycopersicella* and other lepidopteran species in tomato has included careful scouting (primarily with pheromone traps from planting to harvesting), cultural control (including plowing under crop residues promptly after harvesting, cleaning drainage ditches and irrigation canals where alternate hosts grow, and establishing a tomato-free period during summer or winter to break the cycle of tomato pinworm reproduction), mating disruption, use of selective insecticides, and biological control [23].

The parasitoid wasp, *Trichogramma pretiosum* is an egg parasitoid of tomato pinworm and it has been found occurring in several Mexican states (Chihuahua, Coahuila, Durango, Nuevo León, Sinaloa, Sonora, Tamaulipas, and Zacatecas) [27]. This parasitoid species has been released in combination with mating disruption [28]. Due to the overuse of insecticide applications, the tomato pinworm has developed resistance to conventional insecticides so that combined use of pheromones, biological control, and selective insecticides has reduced damage and number of insecticide applications [23].

IPM in cruciferous [the diamondback moth, *Plutella xylostella*]: effective cultural control methods included plowing to eliminate crop residue, and rotation with nonhost crops, careful inspection of nursery plants for diamondback moth eggs and larvae helped to prevent accidental introduction of diamondback moth into the field [29]. In addition, biological control has showed to have an important impact on the control of the diamondback moth, including use of native parasitoid species and the introduction of effective exotic species. In Puebla, a last-instar-parasitoid of diamondback moth, *Diadegma insulare*, has been found parasitizing 46.7% of *Plutella xylostella* larvae in cauliflower [30].

IPM of fruit flies [*Ceratitis capitata*]: according to Mota-Sánchez et al. [26], success of IPM of fruit flies relies on the following crucial steps:

a. Early detection and identification.

b. Reduction of the population by using cultural control, application of selective baits: adult fruit flies are monitored using glass McPhail traps [31] baited with hydrolyzed protein at a density of one to five traps per hectare depending on the species. Fruit sampling is complementary to the trapping and is useful for the detection of larvae. Fruit sampling starts as soon as the orchards and areas outside of the orchards (fruit trees in yards of houses or other hosts in noncommercial areas) have fruits big enough to be infested by fruit flies.

c. Production and release of parasitoids and sterile fruit flies.

d. Strict limitations on fruit movement out of infested areas.

Apart from the strategies for pest control, some other aspects have contributed to the success of IPM programs in Mexico (**Figure 2**).

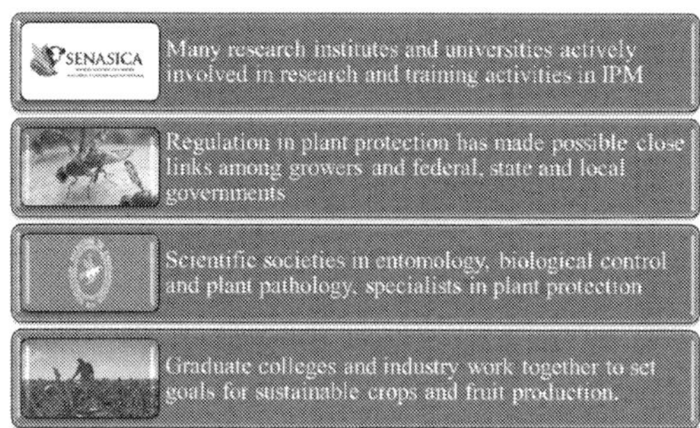

Figure 2. Factors contributing to the success of IPM in Mexico.

However, Mexico still face challenges as some poor farmers cannot afford to implement IPM. Mexico is a country of contrasts where 50 million people live in poverty including poor farmers. Some government programs have been dedicated to improve the conditions of poor people in the country, however, is not an easy problem to solve.

4.5. Colombia

In Colombia, the production of passion fruit (*Passiflora* spp.) is mainly in hands of small farmers. Being cultivated over 8000 hectares, *Dasiops inedulis* (Díptera: Lonchaeidae) is a key pest of passion fruit crop, but there is little information regarding their biology, ecology, and management. Local producers have large production losses due to pests, due to limited knowledge to manage them properly, facing difficulty in positioning their products in the market.

In 2008, Centro Internacional de Agricultura Tropical (CIAT) researchers worked together with local universities and farmer associations to develop a sustainable pest management package for *Dasiops inedulis*. This work allowed farmers to increase their IPM package at the field level. Field surveys conducted from 2008 to 2010 in the main fruit producing regions provided information about the pest population dynamics and geographic patterns of infestation [32]. Then, a national survey of farmers was conducted to get an idea of agroecological behavior management and local knowledge of the farmers. Apart from the common use of insecticide applications based on the calendar of application, they experimented extensively with the farmers the use of inexpensive bait traps. By using participatory practices in five agricultural communities, the farmers realized that some of the new management practices were much more effective and less expensive than current practices of pesticide application [33].

4.6. Peru

IPM in Peru began in the mid-1950s in response to problems caused by the use of organochlorines on crops such as cotton, citrus, olives, and sugarcane [34]. In 1971, graduate programs (MSc level) in entomology and plant pathology were initiated at the National Agrarian University 'La Molina'. In recent years, the Government of Peru has reinitiated technical assistance to farmers through special programs that included the extension of IPM. These programs include Modules of Technical Assistance, coordinated by INIA (Instituto Nacional de Innovación Agraria) at the national level, which have the plant clinics as the diagnostic component, PRONAMACHCS (Programa Nacional de Manejo de Cuencas Hidrográficas y Conservación de Suelos; it is a national program for the management of soils and watersheds), and SENASA (Servicio Nacional de Sanidad Agraria; it is the national service for plant and animal health) [34].

In Peru, most of the vegetable species are usually cultivated in smallholder farms; hence, agricultural production is characterized by lower productivity due to limited availability of good quality seeds and pest problems, besides lack of selected varieties adapted to the agroecosystem [35]. Moreover, most of the farmers do not recognize neither pest species nor beneficial organisms, making insecticide/fungicide applications when is not necessary [35].

All these factors highlight the need to establish an education program for farmers to be trained in sustainable pest management. Saldaña et al. [36] proposed an IPM program for industrialized tomato to manage populations of the two most important pests (*Tuta absoluta* and *Bemisia* spp.) in Barranca, Lima (**Figure 3**). This proposal was based on the pest evaluation strategy, action thresholds, and the application of different control methods, including the establishment of planting dates (legal control), optimization of farming practices (cultural control), installation of light and pheromone traps (ethological control), and removal of virosic plants (mechanical control), maintenance of natural enemies populations (biological control), and selective application of pesticides (chemical control).

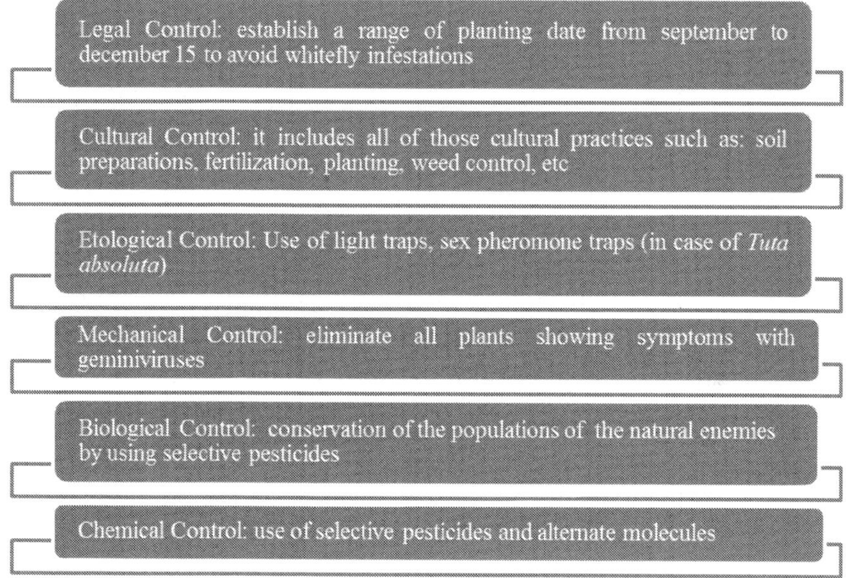

Figure 3. IPM program proposed for pest control in industrialized tomato in Peru (from: [36]).

As a first step, authors developed a methodology to evaluate the specific characteristics of the agricultural ecosystem to determine pest incidence on different phenological stages and establish thresholds to take more efficient control measures. The pest evaluation methodology developed by Sarmiento and Sánchez [37], consists in considering 5 ha as a unit of evaluation which is divided into five subunits. In each subunit, five plants are sampled (four shoots, one leaflet from basal and middle strata, four inflorescences, one twig, and four fruits along 2 m in a furrow).

5. IPM in Latin America: status and challenges

As stated by Rodríguez and Niemeyer [9], IPM research and promotion have responded, in one hand, to food security, which is devoted to the protection of a subsistence crop mainly focused on smallholder peasants, and on the other hand, exports which try to fulfil the requirements of foreign markets and are concentrated in larger producers.

Although research and field-level implementation of IPM has been most successful in the United States and Europe, IPM has made significant progress in developing countries, but focused generally on large-scale rather than small, subsistence farms [38]. According to Rodríguez and Niemeyer [9], government programs and subsidies in developing countries have been concentrated on medium and large farmers since they are able to hire personnel to develop research or to create links with external institutions. Thus, in some countries, such as Chile, there are grant funds available for agricultural research and innovation projects incorporating IPM practices involving partnerships with private firms under a commitment to transfer the results to potential users. Given the requirements for partnerships, the program is not easily available for small farmers, and most research is guided by the specific needs of larger export companies. However, increasingly, scientists, policy makers, and donor agencies in developing countries are turning their attention to small farmers.

Some farmers have benefited greatly from introduced technologies in major production areas in Latin America as many of the new crop technologies have increased crop yield and also their commodity crops can be sent to market [39]. Conversely, those farmers poorly served by markets or have not been reached by modernization packages, the technologies, and practices have failed to generate significant benefits in crop protection systems [40].

The media and public agricultural extension have played a crucial role in introducing the new technologies and good agricultural practices to farmers, however; there has been little investment in farmer education so that they are able to expand their capabilities to understand, innovate, and adapt to the changing context [39]. Although more effort to expand farmers' capabilities to improve production and productivity have been made, agricultural development programs have been unsuccessful because they failed to educate farmers on the sustainable management of variable agroecosystems and to cope with the changes in marketing demands arising from globalizing food and commodity trade [39, 41].

As stated by van den Berg and Jiggins [39], the role of the new generations of farmers has reduced to be simple technology clients, leading them to lose much of the indigenous agricultural knowledge and skills, and in the integrity of the social organization in which indigenous innovation capacity is embedded.

Thereby, the challenge then would be focused to capacitate the millions of small farmers to deal with pest and become experts in decentralized pest management through practical, field-based learning methods.

6. Plantwise helping small farmers to produce in a sustainable way

In some areas, up to 70% of food is lost before it can be consumed. This problem is exacerbated by international trade, intensified production, and climate change altering and accelerating the spread of plant pests. Clearly there is an opportunity to lose less and feed more by improving control of such pest problems, particularly in the developing world [42, 43].

Plantwise (www.plantwise.org), an innovative global program, led by CABI, aims to contribute to increased food security, alleviated poverty, and improved livelihoods by enabling male

http://dx.doi.org/10.5772/64098

and female farmers around the world to lose less, produce more, and improve the quality of their crops. Working in close partnership with relevant actors, Plantwise strengthens national plant health systems from within, enabling countries to provide farmers with the knowledge they need to lose less and feed more [44].

The Plantwise approach is based on three interlinked components:

1. An evergrowing network of locally-run **plant clinics**, where farmers can find advice to manage and prevent crop problems. Agricultural advisory staff is trained to identify any problem on any crop brought to the clinics, and provide appropriate recommendations guided by national and international best practice standards.

2. Improved information flows between everyone whose work supports farmers (e.g., extension, research, input suppliers, and regulators). Collaboration within national **plant health systems** enables these actors to be more effective in their work to improve plant health, with concrete benefits for farmers.

3. The Plantwise **knowledge bank**, a database with online and offline resources for pest diagnostic and advisory services, provides both locally relevant, comprehensive plant health information for everyone and a platform for collaboration and information sharing between plant health stakeholders.

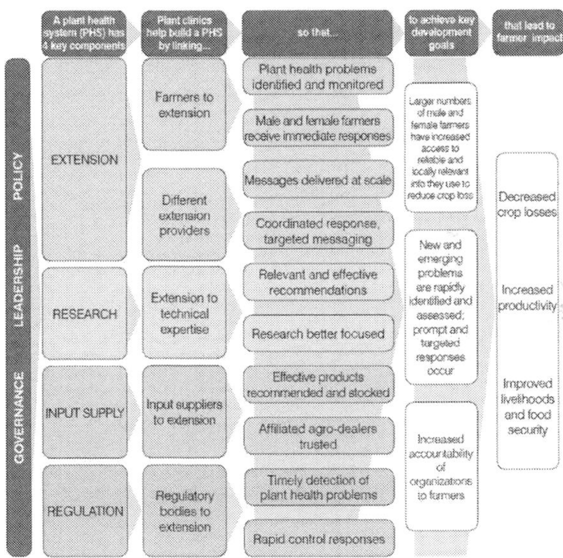

Figure 4. Plantwise theory of change (from: [46]).

In the Plantwise knowledge bank, plant clinic records are collated and analysed to support the quality of advice given to farmers and inform decision-making. By putting knowledge into the hands of smallholder farmers, Plantwise is able not only to help them lose less and feed more but also to gather data which can assist all stakeholders in the plant health system-from research, agro-input supply, extension and policy-making. Most importantly, Plantwise is a development program which cooperates with a number of international and national organizations working to remove constraints to agricultural productivity. Countries are now using plant clinics and Knowledge bank resources to improve national vigilance against pest outbreaks [45].

The key premise of the Plantwise Theory of Change is that plant health systems function to reduce crop losses and promote plant health (**Figure 4**). Plantwise defines a plant health system by four key components: (1) extension, which delivers available knowledge intended to improve plant health; (2) research, which develops new knowledge about plant health and is often linked to higher level education; (3) input suppliers, who deliver knowledge and physical inputs such as seeds, biological and other crop protection products, and fertilizers; and (4) regulation, which regulates sale and use of agricultural inputs, protects countries from new and emerging pests (invasive species included), and regulates produce export requirements.

The Plantwise approach develops sustainable mechanisms to deliver better plant health services that address farmer needs and improve output, including (1) improving advisory services based on plant clinics and complementary extension approaches and delivering effective responses to any plant health problem affecting any crop; (2) improving regulatory systems so that plant health problems are detected early and advisory staff on the ground are able to communicate appropriate mitigation measures to farmers before the problems become devastating; (3) stimulating research that supports farmers' needs; and (4) improving input supply ensuring provision of appropriate, legitimate, and effective goods [46].

The Plantwise programme encourages extension officers to offer plant health management advice to farmers guided by the principles of integrated pest management (IPM), looking forward to increase the sustainability of the production system [46].

Author details

Yelitza Colmenárez[1*], Carlos Vásquez[2], Natália Corniani[3] and Javier Franco[4]

*Address all correspondence to: y.colmenarez@cabi.org

1 UNESP-FEPAF-Fazenda Experimental Lageado, Botucatu - SP, Brazil

2 Universidad Técnica de Ambato, Facultad de Ciencias Agropecuarias, Campus Querochaca, Cevallos, Ecuador

3 CABI South America, Botucatu, Brazil

4 Proinpa Foundation, Cochabamba, Bolívia

References

[1] Fitt G, Wilson L. Integrated pest management for sustainable agriculture. In: Abrol DP, Shankar U, editors. Integrated Pest Management. Wallingford, Oxfordshire, England: CABI Publishing; 2012. p. 27–40.

[2] Keating BA, Carberry PS, Bindraban PS, Asseng S, Meinke H, Dixon J. Eco-efficient agriculture: concepts, challenges and opportunities. Crop Science 2010;50:109-119.

[3] Yudulmen M, Ratta A, Nygaard D. Pest management and Food Production. International Food Policy Research Institute Discussion Paper 25. Washington, DC: IFPRI; 1998.

[4] Gill HK, Garg H. Pesticide: environmental impacts and management strategies. In: Solenski S, Larramenday ML, editors. Pesticides Toxic Effects. Rijeka, Croatia: Intech; 2014. p. 187–230.

[5] Ster VM, Smith RF, van den Bosch R, Hagen KS. The integration of chemical and biological control of the spotted alfalfa aphid (the integrated control concept). Hilgardia 1959;29:81-101.

[6] Naranjo SE, Ellsworth PC. Fifty years of the integrated control concept: moving the model and implementation forward in Arizona. Pest Management Science 2009;65:1267-1286.

[7] Ellsworth PC, Martinez-Carrillo JL. IPM for *Bemisia tabacia*: a case study from North America. Crop Protection 2001;20:853-869.

[8] Kogan M. Integrated pest management: historical perspectives and contemporary developments. Annual Review of Entomology 1998;43:243-270.

[9] Rodríguez LC, Niemeyer HM. Integrated pest management, semiochemicals and microbial pest-control agents in Latin American agriculture. Crop Protection 2005;24:615-623.

[10] Coble HD, Ortman EE. The USA national IPM road map. In: Radcliffe EB, Hutchinson WD, Cancelado RE, editors. Integrated Pest Management: Concepts, Tactics, Strategies and Case Studies. Cambridge: Cambridge University Press; 2009. p. 471-478.

[11] Flint ML, van den Bosch R. Introduction to Integrated Pest Management. Netherlands: Springer Science and Business Media; 2012. 260 p.

[12] Bird GW. Role of integrated pest management and sustainable development. In: Maredia KM, Dakouo D, Mota-Sanchez D, editors. Integrated Pest Management in the Global Arena. Wallingford, Oxfordshire, England: CABI Publishing; 2003. p. 73–86.

[13] Peshin R. Evaluation of the dissemination of insecticide resistance management programme in cotton in Punjab [dissertation]. Ludhiana, 2005.

[14] Mumford JD, Norton GA. Economics of decision making in pest management. Annual Review of Entomology 1984;29:157-174.

[15] Orr D. Biological control and integrated pest management. In: Peshin R, Dhawan AK, editors. Integrated Pest Management: Innovation-Development Process. Netherlands: Springer Science + Business Media; 2009. p. 207–240.

[16] O'neil RJ, Yaninek JS, Landis DA, Orr DB. Biological control and integrated pest management. In: Maredia KM, Dakouo D, Mota-Sanchez D, editors. Integrated Pest Management in the Global Arena. Wallingford, Oxfordshire, England: CABI Publishing; 2003. p. 19–30.

[17] Tauber MJ, Hoy MA, Herzog DC. Biological control in agricultural IPM systems: a brief overview of the current status and future prospects. In: Hoy MA, Herzog DC, editors. Biological control in Agricultural IPM Systems. Orlando, FL, USA: Academic Press INC.; 1985. p. 9-12.

[18] Murphy ST, Briscoe BR. The red palm weevil as an alien invasive: biology and the prospects for biological control as a component of IPM. Biocontrol News and Information. 1999;20(1):35-46.

[19] Mazza G, Francardi V, Simoni S, Benvenuti C, Cervo R, Faleiro JR, Llácer E, Longo S, Nannelli R, Tarasco E, Roversi PF. An overview on the natural enemies of *Rhynchophorus* palm weevils, with focus on *R. ferrugineus*. 2014;77:83–92.

[20] Bajwa WI, Kogan M. Integrated pest management adoptions by global community. In: Maredia KM, Dakouo D, Mota-Sanchez D, editors. Integrated Pest Management in the Global Arena. Wallingford, Oxfordshire, England: CABI Publishing; 2003. p. 97–107.

[21] Carmona D, Huarte M, Arias G, López A, Vincini AM, Álvarez HA, Manetti P, Capezio S, Chávez E, Torres M, Eyherabide J, Mantecón J, Cichón L, Fernández D. Integrated pest management in Argentina. In: Maredia KM, Dakouo D, Mota-Sanchez D, editors. Integrated Pest Management in the Global Arena. Wallingford, Oxfordshire, England: CABI Publishing; 2003. p. 313–326.

[22] Parra JR. Manejo integrado de pragas. In: Paterniani E, editor. Agricultura Brasileira e a Pesquisa Agropecuária. Brasília: Embrapa Comunicação para Transferência de Tecnologia; 2000. p. 92-105.

[23] Hoffmann-Campo CB, Oliveira LJ, Moscardi F, Gazzoni DL, Corrêa-Ferreira BS, Lorini IA, Borges M, Panizzi AR, Sosa-Gomez DR, Corso IC. Integrated Pest Management in Brazil. In: Maredia KM, Dakouo D, Mota-Sanchez D, editors. Integrated Pest Management in the Global Arena. Wallingford, Oxfordshire, England: CABI Publishing; 2003. p. 285–300.

[24] Emilio Farias Falcones. Manejo Integrado de Plagas en Cacao en el Ecuador. Available from: http://elproductor.com/2012/07/09/manejo-integrado-de-plagas-en-cacao-utilizando-new-bt-2x [accessed: 30/03/2016].

[25] Mendoza J, Gualle D, Gómez P, Ayora A, Martínez I, Cabezas C. Progresos en el manejo de plagas en caña de azúcar en Ecuador [Internet]. Available from: http://www.aeta.org.ec/2do congreso cana/art_campo/MENDOZA cana.pdf [accessed: 30/03/2016].

[26] Mota-Sanchez D, Santos-González F, Alvarado-Rodríguez B, Díaz-Gómez O, Bravo-Mojica H, Santiago-Martínez G, Bujanos R. Integrated pest management in Mexico. In: Maredia KM, Dakouo D, Mota-Sanchez D, editors. . Integrated Pest Management in the Global Arena. Wallingford, Oxfordshire, England: CABI Publishing; 2003. p. 273–284.

[27] García-González F, Mercado-Hernández R, González-Hernández A, Ramírez-Delgado M. Especies nativas de *Trichogramma* (Hymenoptera: Trichogrammatidae) colectadas en cultivos agrícolas del Norte de México. Revista Chapingo Serie Ciencias Forestales y del Ambiente. 2011;17:173-181.

[28] van Driesche R, Hoddle M, Center T. Control of pests and weeds: an introduction to biological control. Blackwell Publishing, USA; 2008. 484 p.

[29] Bujanos MR, Marin JA, Galvan CF, Byerly MKF. Manejo Integrado de la Palomilla dorso de diamante (*Plutella xylostella* (L.) Lepidoptera: Yponomeutidae) en el Bajío México. INIFAP Asociación de Procesadores de Frutas y Vegetales en General; 1993. 36 p.

[30] Tovar-Hernández H, Bautista-Martínez N, Vera-Graziano J, Suárez-Vargas A, Ramírez-Alarcón S. Fluctuación poblacional y parasitismo de larvas de *Capitarsia decolora* (Guenée), *Plutella xylostela* L. y *Trichoplusia ni* (Hübner) (Lepidoptera) en *Brassica oleracea* L. Acta Zoologica Mexicana. 2007;23(2):183–196.

[31] Steyskal GC. History and use of the McPhail trap. The Florida Entomologist. 1977,60:11–16.

[32] Rengifo JA, Garcia JG, Rodriguez JF, Wyckhuys KAG. Host status of purple passion fruit for the Mediterranean fruit fly (Diptera: Tephritidae). Florida Entomologist 2011;94:91–96.

[33] Wyckhuys KAG, Lopez F, Rojas M, Ocampo JA. Do farm surroundings and local infestation pressure relate to pest management in three cultivated *Passiflora* species in Colombia? International Journal of Pest Management. 2011;57:1–10.

[34] Palacios-Lazo M, Lizárraga-Travaglini A, Velásquez-Ochoa R, Carranza-Hernández E, Segovia I. Integrated pest management in Peru. In: Maredia KM, Dakouo D, Mota-Sanchez D, editors. Integrated Pest Management in the Global Arena. Wallingford, Oxfordshire, England: CABI Publishing; 2003. p. 301–312.

[35] Cañedo V, Alfaro A, Kroschel J. Manejo Integrado de Plagas: insectos en hortalizas, principios y referencias técnicas para la Sierra Central de Perú. Centro Internacional de la Papa; 2011. 54 p.

[36] Saldaña C, Sarmiento J, Sánchez G. Manejo Integrado de plagas en el cultivo industrial de tomate (*Lycopersicum esculentum* Mill.) en el Valle de Barranca, Lima, Perú. Revista Peruana de Entomología. 2033;43:153–157.

[37] Sarmiento J, Sánchez G, editors. Evaluación de Insectos. Lima: Universidad Nacional Agraria; 2000. 117 p.

[38] Goodell G. Challenges to international pest management research and extension in the third world: Do we really want IPM to work? Bulletin of the Entomological Society of America. 1984;30(3):18–26.

[39] van den Berg H, Jiggins J. Investing in farmers – the impacts of farmer field schools in relation to integrated pest management. World Development. 2007;35(4):663–686.

[40] Inter-Academy Council. Realizing the promise and potential of African Agriculture: science and technology strategies for improving agricultural productivity and food security in Africa. Amsterdam: Inter-Academy Council; 2004.

[41] Reardon T, Berdegué JA. The rapid rise of supermarkets in Latin America: challenges and opportunities for development. Development Policy Review. 2002;20:371–388.

[42] Oerke EC. Crop losses to pests. Journal of Agricultural Science 2006;144:31–43.

[43] Finegold C, Oronje M, Leach MC, Karanja T, Chege F, Hobbs SLA. Plantwise knowledge bank: building sustainable data and information processes to support plant clinics in Kenya. Agricultural Information Worldwide. 2014;6:96–101.

[44] Romney DR. Plantwise: putting innovation systems principles into practice. Agriculture for Development. 2013;18:27–31.

[45] Kuhlmann U. Plantwise: a global alliance for plant health support. International Agriculture in a changing world: good news from the field; Bern. 2014. p. 23.

[46] Plantwise. Plantwise Strategy 2015–2020 [Internet]. Available from: https://www.plantwise.org/Uploads/Plantwise/Plantwise%20Strategy%202015%202020.pdf [accessed: 2/02/2016].

Chapter 2

Pest Control in Organic Systems

Vasile Stoleru and Vicenzo Michele Sellitto

Additional information is available at the end of the chapter

http://dx.doi.org/10.5772/64457

Abstract

Conventional agriculture techniques applied in the latest decades have had undesired consequences on the environmental sustainability, carried out to the soil erosion, the degradation of the ecological system, changing the balance between beneficial and harmful pests, and contamination of soil, water, and agricultural products by heavy metals and pesticides. Thus, in organic agriculture, using synthetic chemicals for pest control is prohibited, assigning to the diversity a major role. The study provides to the reader many important practical data, judiciously documented, which are useful for the researchers and farmers from the world. Pest control in organic agriculture can be obtained through prevention and curative measure, but modern agriculture must be focused on the prevention.

Keywords: organic agriculture, pest control, preventive and curative methods

1. Introduction

Organic agriculture (OA) farming aims to achieve sustainable, diversified, and balanced systems, with the purpose of protecting the environment for present and future generations. In the same way, OA provides on the food market, products of a certain nutritional quality, suitable in terms of lower contaminants.

The organic product is governed by some well defined principles, aimed at ensuring environmental and crop sustainability.

1.1. Circumstances of pest control in organic systems

Being a type of sustainable agriculture the purpose of OA can be expressed by a mini–max function, maximizing production and minimizing the negative agricultural activities on the environment [1].

OA stimulates the activity of useful microorganisms, flora and fauna. Soils under crops are increasingly lifeless and infested with weeds, diseases, as well as pests. This situation is determined by current agricultural practices that excel in monoculture and short crop rotations, of 2–3 years, much delayed and bad quality soil tillage and plant care, burning plant debris, etc.

Biodiversity management. The soil's biological resources are vital to the economic and social development of all humanity. That is why, it is more and more frequently recognized that biological diversity is universal asset, of inestimable value for future generations. Biological (ecologic, organic) agriculture generally uses a greater number of cultivated species, to explore their suitability and ecological plasticity. Non-using synthetic herbicides, and instead using milder solutions for weed destruction, ensures the coexistence of weeds together with the crop.

Protecting the natural landscape. Elevation diversity, as well as flora and fauna variability, is inseparable to the applied vegetable growing systems, the most aggressive ones being of the intensive type, often causing deterioration.

Many cultivation techniques applied in the past decades have had undesired consequences on the environment, contributing to soil erosion, the degradation of the ecological system, contamination of ground water and crops with pesticides and nitrates.

Organic agriculture aims to preserve the environment unaltered, using organic fertilizers and also less soluble mineral fertilizers, organic fertilizers, such as composts and green fertilizers, avoiding to use products that can have harmful effects [2].

The use of synthetic herbicides and pesticides are prohibited, and only products that are harmless for the plant are allowed, products based on simple minerals (Cu, S, Na, silicate, etc.) or plant extracts (pyrethrum), including the application of physical (thermal) methods.

In organic agriculture, the emphasis is laid on the quality of human intervention over nature, which is non-aggressive, compared to conventional agriculture.

1.2. Standards and regulations regarding organic farming

After 2010, OA can be considered a period of consolidation for standards and the regulations, which aimed and still aims to facilitate international trade with organic products in order to reduce legislative gaps which exist among the various certification types, such as the EC Regulations [3, 4], the USA (NOP), Australia (AS 6000-2009), Japan (JAS), and Switzerland (Bio Swiss). Thus, the EC Regulation of organic agriculture [3] has been improved in the least years with new regulations, targeting aquaculture and organic wine production.

The number of the certification bodies, in 2013, was at 569, increasing from 2010 when there were 532. Most certification bodies are found in the European Union, Japan, the United States of America, South Korea, China, Canada, India, and Brazil [5].

Organic farming (biological, ecological) is currently one of the most dynamic forms of agriculture. This affirmation is mainly supported by the expansion of agricultural areas, currently occupying 40.2 of the surface in Oceania, 26.6 in Europe, and 15.3% in Latin America. There are also cases of countries, such as Argentina, Spain, and USA, in which the area increased in 2013 compared to 2011 with over 185,000 ha.

Around the world, at the end of 2013, the organically certified area covered more than 78 million ha. Organically certified agricultural areas covered over 43 million ha (1% from total arable land), including the same land under its conversion period, but excluded wild collection and aquaculture. From these data, it appears that the organically administered surface has had a growth rate of over 14.94% compared with 2012 (approx. 37.4 million ha). Europe and Oceania recorded the fastest land expansion rhythm in 2013, compared to 2011, which shows that the expansion of the areas is supported by an intensive marketing of organic products [5].

Compared to 2012, the organically certified area in the world increased by over 5.6 million ha, which means a growth rate of the arable production from the total agricultural area of 0.1%.

At the end of 2013, the situation of the organic agricultural area distributed on categories of land use highlighted that 63% was permanent grassland, 18% was arable land (cereals, green fodder, oilseed, vegetable, and protein crops), and 7% was permanent crop (coffee, olives, nuts, grapes, and cocoa) and the rest with other crops [5].

Of course, in some countries, the conversion areas or the cultivated ones are decreasing, especially due to legislation and government support, which differ from country to country (UK).

Global sales of organic food and drinks reached more 72 billion dollars at the end of 2013. Compared to 2009, this sector revenue increased almost five times. Europe and North America made a big contribution to cover these specific sectors. Asia, Latin America, and Africa have become really important producers of organic crops for this market. About 43% from this market is covered by the United States followed by Europe at percent 40% [5].

In 2013, the countries with the largest organic markets were the USA (24.3 billion €), Germany (7.6 billion €), and France (4.4 billion €) [5].

2. Pest control measures

Organic farming (OF) is a system-based agricultural production system working with rather than against natural systems [2].

The major differences that have been made in terms of technology between organic and conventional cultivation of plants are as follows: soil fertility, weeds, pathogens, and pest control.

Pest control in organic agriculture can be obtained through prevention and curative measure but must be focused on the preventive infestation of pests [2]. Measures to prevent infestation by pests refers to: phytosanitary quarantine (special for seed and planting materials used for establishing crops); monitoring pest infestation (used in general agro-expert stations or traps); choice of cultivars according to the criterion of resistance and ecological plasticity; seed conditioning; destruction of problematic weeds; solarization; and hygienic conditions.

2.1. Prevention pests in an organic system

The fundamental principle of controlling pests in organic systems (OS) should consider the mechanism of adjusting its biocenosis (total community of organisms from o biotope), through the correlation and interdependence between the cultivated species, pathogens, weeds, pests, technology, and the environment. Protecting plants from pests and diseases probably has the greatest impact on achieving an organic vegetable crop, due to the very large spectrum of pathogens and pests from these crops. The first major attempt to reduce chemical treatments took place even before 1970, when the concept of integrated control was promoted [6, 7]. According to this concept, all technical methods are allowed to maintain the populations of pests and pathogens under a certain degree of impairment, which does not affect the yields from an economic point of view.

This concept is approved by the International Organization for Biological Control (IOBC), but first of all natural factors must be used, together with other methods appropriate for the economic, ecological, and toxicological requirements [8].

In organic farming, the principles of the integrated pest control are perfectly applicable in substantializing the mechanisms for fighting pests, diseases, but most chemical means are forbidden; instead, new unconventional methods have been used, like some biodynamic preparations.

The strong attack of some pests may be favored by some technical mistakes, in general, or mistakes in the environmental context such as the following: improper choice of the place of culture; using seeds or plants that are weakly developed; mistakes in crop association; practicing monocultures without using proper crop rotation; incorrectly executed soil tillage; unilateral or excessive fertilization, without organic fertilizers; insufficient fertilization; extreme weather conditions; and improper choice of the sowing period [1, 9].

2.1.1. Phytosanitary quarantine

The quarantine is a complex of preventive measures taken to stop the penetration of diseases, pests, or weeds from other countries and to limit their spread. Overall, export products between countries shall be binding accompanied by a phytosanitary document certifying that the seeds or agricultural materials for setting up the crop (seeds, cuttings, tubers, bulbs, seedlings, shrubs, or trees) are free from pest quarantine.

There are numerous species (mites, insects), generally in polyphagus that are considered extremely dangerous and huge efforts have been made to limit their expansion, for example:

Leptinotarsa decemlineata (Colorado beetle), *Tetranychus urticae* (red spider mite), *Myzus persicae* (green peach aphid), *Bemisia tabaci* (silverleaf whitefly), *Trialeurodes vaporariorum* (greenhouse whitefly), *Liriomyza trifolii* (leaf miner flies), *Tuta absoluta* (tomato leaf miner), *Spodoptera litura* (Oriental leaf worm moth), *Frankliniella intonsa red* (red thrips), *Diabrotica virgifera virgifera* (western corn rootworm), or others [10–12].

2.1.2. *Maintenance of biodiversity*

Synthetic pesticides are not permitted in organic farming which serves to preserve and enhance biodiversity within the system. Natural enemies of pest species are therefore able to thrive, exerting control on pest populations. Conservation and improvement of natural features of the landscape, such as hedgerows and ponds and the construction of beetle banks and sown flower strips, have also enabled communities of predators to flourish.

In agriculture, in general, farmers work with biological organisms, which behave differently under the action of nature's biotic or abiotic factors [13].

The pests are very adaptive to the changes of production systems, especially from the transfer from conventional to organic farms (in conversion).

In OA, pest problems are influenced by three major components of farming systems, such as: crop species and cultivar, agro-ecosystem structure, and technology production (**Figures 1** and **2**).

Figure 1. Management of land for organic agriculture (photograph by Stoleru Vasile).

Researchers developed flowering strips that are tailored to requirements of the specific complex of natural enemies within a cropping system. So, any experiments identified selective plant species that would improve the longevity and parasitization rate of the parasitoid wasps (*Microplitis mediator, Diadegma fenestrale,* and *D. semiclausum*) on the *Mamestra brassicae.*

Figure 2. Sea buckthorn hedge on an organic farm (photograph by Stoleru Vasile).

Comparing the effects of floral and extrafloral nectar of different plants, beneficial effects of *Fagopyrum esculentum* (floral nectar), *Centaurea cyanus* (floral and extrafloral nectar), and non-flowering *Vicia sativa* (extrafloral nectar) on parasitoids were found. Extensive plant screening is essential to achieve plant selectivity and to maximize biological control. *F. esculentum, C. cyanus* and *V. sativa* are recommended as selective plant species to enhance parasitoids of *M. brassicae* [14].

2.1.3. Selection of cultivars according to the resistance and ecological plasticity criteria

The cultivar is perhaps the most important factor that productivity and quality depend on. Because of its biological and technological potential, it will be expressed in terms of appropriate measures [15].

In order to choose the most suitable cultivar for OA, the farmer should take into account main criteria: consumer preferences regarding appearance, taste [2], etc.; climate and soil conditions, adaptation to extreme environmental conditions; extreme temperatures, the length of the photoperiod, tolerance to high concentrations of salts, and economic use of fertilizers; resistance or tolerance to diseases and pests; cultivation technology (field, greenhouse, tunnels, time of sowing, planting and the harvesting period, irrigated regime or less, mechanization) [16]; and product destination: fresh consumption and industrialization (canning, freezing, dehydration, etc.);

A cultivar cannot meet all these requirements, but, depending on the destination of the products and both the consumers' requirement and farmers' preferences, the most suitable biological material will be chosen under the given conditions [17].

http://dx.doi.org/10.5772/64457

There are very different requirements from the growers regarding variety characteristics, depending on the size of the surfaces and the destination of the products. Thus, for small gardens, created by amateurs for their own consumption, large fruit species can be cultivated, as they are more sensitive to transport and storage. OA can be used as varieties, hybrids, local populations, and clones [3, 4], but not accepted genetically modified organisms.

Choosing varieties and hybrids with resistance to pathogens and pests is necessary both for protected crops and for early field crops, because the investment is often large, so risks and loss must be eliminated [18–20].

For many crops (tomatoes, cucumbers, eggplants, bushes, or trees) grafted method may be used that causes plant vigor and thus resistance to nematode (**Figure 3**).

Figure 3. Fado hybrid grafted on the Rezistar rootstock for an attack on nematodes (photograph by Stoleru Vasile).

In **Table 1** are presented any cultivars with resistance or tolerance to the attack of different pests, especially for nematode control, in temperate climate conditions.

Recent research on the outside cabbage crop in the temperate climate highlighted, Timpurie de Vidra cultivar (cv) of early cabbage is most resistant to the cabbage fly (8.5% degree of attack) in comparison with the Golden acre cv., where the degree of attack was 14.2%, during two study years [21].

The reaction of cultivars resistant to pests and the nematode default may be determined by its presence in the plant silica [22], iron [23] genes that provide resistance [18, 24, 25], or protein presence in bean or cowpea [23, 26, 27].

Species	Cultivar	Pest resistant or tolerance
Tomato	Getina F1 Gloria F1, Splendid F1, Solara F1, Nemarom F1	*Meloidogyne incognita Chitw., Meloidogyne hapla Chitw.*
Dianthus	Sooty	*Meloidogyne arenaria Neal.,*
Pineapple	Turiacu	*Meloidogyne arenaria Neal.,*
Zucchini	Amalthee	*Meloidogyne* spp.
Cucumber	Dasher II	*Meloidogyne* spp.
Soybean	Huasteca 300	*Tamaulipas state*

Table 1. Varieties created with resistance or tolerance to various pests.

2.1.4. Seed conditioning

Numerous pests, especially from the coleopteran order, can be found between the seeds or inside them during sowing, as they feed within their endosperm, endangering seed germination or weakening the newly sprouted plant [8]. The larvae and adults of nematodes (*Ditylenchus dipsaci, Tylenchorhynchus cylindricus*) attack both the garlic and onion bulbs but also the roots of the vegetables, making the plant die dry [28, 29].

2.1.5. Crop rotation

Effective crop rotations are fundamental to pest control in OS. Correct rotations provide an obstacle to the pest life cycles by removing host crops for prolonged periods of time. They also help in supporting a more diverse and stable agro-ecosystem to assist with natural pest suppression.

In areas where the climate permits, two or three crops can be grown during on the year on the same area, both in greenhouses or tunnels. From this point of view, it must be considered as species that succeed have no common pests (**Table 2**).

No.	Crop	Sowing period	Planting	End of the crop
1	Lettuce, anticipated	20 VIII–10 IX	20 IX–10 X	15–30 III
2	Sweet pepper	10–15 I	1–15 IV	20–30 IX
3	Green onion	25–30 IV	1–15 X	25 III–5 IV

Table 2. Plot design for successive crops in greenhouse/tunnel.

For the outdoor crops, in **Table 3**, some design rotation successive crops are presented. For these designs bear in mind that the crops that are grown on the same land area must belong to the same botanical family, have no common diseases and pests, and have different growing seasons.

Type of crop	Crop	Sowing period	Planting	End of the crop
Vegetable design (1)	Radish	10–20 III	–	10–15 IV
	Tomato	10–15 II	15–20 IV	25–30 IX
Vegetable design (2)	Peas pods	1–15 III	–	10–20 VI
	Late cabbage	10–15 V	20–30 VI	15–20 X
Mixed crop design (1)	Barley	20–30 IX	–	10–15 VI
	Cauliflower	10–15 V	25–30 VI	1–10 X
Mixed crop design (2)	Wheat	10–20 X	–	10–20 VI
	Cucumber	1–10 VII	–	1–5 X

Table 3. Plot design for successive crops in the field.

2.1.6. Crop monitoring

Monitoring insects is fundamental in organic farming systems (OFS). Correct identification of insects and insect biology knowledge when they colonize crops is one of the main activities of management decisions that lead to optimal moment. This can be done by simply checking the crop (aphids, spider mites) or by using pheromone traps (thrips, cydia, white fly, rose fly, carrot fly and cabbage moth).

Figure 4. (a) Eggplant leaf affected by *Thrips tabaci* (photograph by Stoleru Vasile). (b) Tomato root affected by *Thrips tabaci* (photograph by Stoleru Vasile).

Pests of agricultural crops can cause damage directly (lower leaf surface, destroying fruit), (**Figure 4a**) or indirectly (gale or gates run for various soil diseases, such as *Rhizoctonia* sp. or *Fusarium* sp.), because many pests performing the biological cycle in soil (**Figure 4b**).

Prognosis and warning are performed by the centers dealing with plant protection, and they establish, at the right moment, the imminent danger of setting off massive pest attacks.

2.1.7. Management practices when it comes to pest control

Cultural activities in organic farming may be considered as specific as crop production practices that implemented in the initial stages of the organic farm plan to reduce the likelihood of insect pest infestation. These measures are based on disrupting the biological cycle of the pest as follows: an unavailable crop to pests in space and time; unacceptable crop to pests by interfering with location; reducing the pest on the crop by natural enemies, etc.

Cultural practices are among the oldest techniques used for pest suppression, and many of the practices used in conventional and organic farming today have their roots in traditional agriculture. Effective deployment of cultural tactics is information intensive; it requires knowledge of pest–crop interactions and about the natural enemies of the pests.

2.1.8. Intercropping system

Intercropping is the practice of growing two or more crops (usually different families) in the same area. Strip cropping is a derivation of intercropping and is the practice of growing two or more crops in alternating strips across a field. Both practices serve to increase biodiversity and make the habitat less suitable for pest development (**Figure 5a, b**).

Figure 5. (a) Intercropping management of the runner bean with maize (photograph by Hamburda Silvia). (b) Intercropping management of the runner bean with sunflower (photograph by Hamburda Silvia).

http://dx.doi.org/10.5772/64457

2.1.9. Tillage management

Much of the pest population from both soil and foliar can be influenced through tillage practices. Tillage systems reduce insect pressure in succeeding crops. Fields are usually tilled in the fall or early spring when many kinds of insects are in the overwintering stage within the soil or in crop residues. Direct destruction of the insect or its overwintering chamber, removal of the protective cover, elimination of food plants, and disruption of the insect life cycle generally kills many of the insects through direct contact, starvation or exposure to predators, and weather.

Crop irrigation by sprinkler reduces the number of pests in crops [30]. Irrigation by culverts reduces the number of galas in the soil and thus causes interruption of the biological cycle of soil insects.

2.1.10. Mulches

Mulch is a layer film of material applied to the soil surface for the following reasons: to conserve moisture, to improve the fertility and health of the soil, to reduce weed growth, and to pressure soil land crop infestation with different pests [31].

Mulch is usually but not exclusively organic in nature (**Figure 6a**). It may be non-biodegradable (e.g., plastic sheeting) or biodegradable (e.g., bark chips). It may be applied to bare soil or around existing plants. Mulch consisting of manure or compost is incorporated naturally into the soil through the activity of worms and other organisms [32].

(a) (b)

Figure 6. (a) Organic cabbage mulched with phacelia (photograph by Stoleru Vasile). (b) Organic cabbage mulched with biodegradable plastic (photograph by Stoleru Vasile).

All mulch types suppress insects in comparison with bare soil. Different colors of plastic have been tested; clear, white, yellow, or aluminum (reflective) colors may provide some additional suppression of aphids and whiteflies [33]. Blue and yellow may bring in more pests. Plastic can be painted the desired color (**Figure 6b**). Before choosing a mulch type, farmers should check with their certifier bodies to see whether the practice is allowable by organic regulations [34].

2.1.11. *Optimum crop health*

The driving force behind the sustainability and environmental preservation derived through organic farming comes through healthy living soil. Microbes in the soil process organic matter to provide a balance of minerals and nutrients which are utilized by plants to achieve healthy, vigor crop growth. When this balance is achieved, the associated health of the crop gives it a heightened ability to withstand pest and disease attack. Good crop husbandry and hygiene also make a significant contribution to the health of the crop and the prevention of pest problems.

2.1.12. *"Host weed" removal*

Numerous dangerous species find favorable conditions for the summer or winter diapause on the spontaneous vegetation from the forest skirt, the borderline of strip ground, roads, or railways or the less cared for agricultural crops. So, the cabbage aphid has as host plant the cole, and the Colorado beetle has as host plant the black nightshade—*Solanum nigrum* [8].

Storing crops in hygienic conditions generally represents an additional source of pest infestation. (e.g., the bean weevil (*Acanthoscelides obsoletus*), pea weevil—*Bruchus pisorum*). They can be fought against either by storing the products in refrigerated storerooms for a certain period of time or by vacuuming the products in a special room [35].

2.2. Curative measures

Curative care or curative measure is the health care given for environmental conditions where a measure is considered achievable, or even possibly so, and directed to this end. Curative care differs from the preventive method, which aims at preventing the appearance of pests, which concentrates on reducing the degree of the attack.

2.2.1. *Physical–mechanical methods*

According to specific regulation (EU 834/2007), in OA, it can be used following measures: thermotherapy, heliotherapy, radiotherapy, ultrasounds, nets, fences, or traps.

Thermotherapy is recommended only if the vegetal remains are highly infested with pests and, as much as possible, after collecting and removing the remains from the cultivated area. In OA according to EU Regulation 834/2007, this method is restrictive and can be applied only in problematic crops. If this is not possible, in situ burning may be used, but only after a thorough investigation of the opportunity of such a measure and registering it in the farm register and announcing the local organization of environmental protection (EU 889/2008).

Heliotherapy. The method is very simple and has been the subject of thorough research studies carried out at the Central Food Technological Research Institute in India [1]. This method consists of exposing the infested seeds to a temperature of 60°C for 10 min [17]. In order to do so, seeds are put in a dark color polyethylene bag with high molecular and density weight, which, at its turn, is tightly covered by another transparent low density polyethylene bag. The entire operation is carried out on a plane surface exposed to sun. The two foils act as a condenser making the temperature inside the seed bag quickly increase leading to the pests' death.

Radiotherapy is used for sterilizing males with the aid of X-rays and gamma radiations. Achieving the dominant lethal mutations has led to obtaining a biological method called autocide.

Figure 7. Ultrasound for pest control (photograph by Stoleru Vasile).

Scientifically, literature mentioned the effects of X-ray irradiation applied on six floriculture insect pests (*Tetranychus urticae, Myzus persicae, Bemisia tabaci, Liriomyza trifolii, Spodoptera litura,* and *Frankliniella intonsa*) placed in the bottom sections of rose and chrysanthemum pots. After irradiation with an X-ray dose of 150 Gy, the development of nymphs and adults of *M. persicae* and eggs, nymphs, and adults of *B. tabaci* was prevented at every position in the pots. *T. urticae* nymphs irradiated at 200 Gy newly emerged adults laid eggs in the bottom section of rose boxes only. *L. trifolii* adults irradiated at 200 Gy were completely inhibited. Radiotherapy method depends on dose of X-ray irradiation, insects, and crops [10].

Other physical or mechanical methods refer to installing various barriers, such as: nets for carrot fly, ultrasounds for soil insects (**Figure 7**), metallic fences for snails (**Figure 8**), layers for aphids and Lepidoptera's insects (**Figure 9**), traps or rollers (carrot fly, thrips) (**Figure 10**), flooding, and crushing the eggs of caterpillars or even the adults.

Figure 8. Metallic fence for protection against the snail (photograph by Stoleru Vasile).

Figure 9. Early crops protected with Agryl P_{17} (photograph by Stoleru Vasile).

Figure 10. Thrips and whitefly plaque applied in tomato crops (photograph by Stoleru Vasile).

Flooding provides better results in fighting against underground pests (mice, moles, crickets, etc.) by flooding their galleries. The impossibility of knowing the exact side of their galleries reduces the method's practical value and limits its use [36].

2.2.2. *Biotechnical methods*

Installing food bait traps. They can consist of parts of plants, fruits, tubercles, or feed and are placed on the ground or in storehouses. After collecting the pests, traps are removed, soaked in boiling water or burnt [31, 37].

Installing pheromone traps. Pheromones are chemical substances secreted and spread outside the body and determine a response only from the individuals of the same species (**Figure 11**). There are multiple types of pheromones, according to the role they fulfill: sexual, alarm, aggregation, path marking, recognition, and social regulation (e.g., ATRAGAM and ATRAPOM are a sexual pheromone used for *Autographa gamma* and *Cydia pomonella*) [8, 38].

Table 4 presents other products that can be applied in organic farming, based on the pheromones.

Figure 11. Attractive traps for pests control in tomato crops (photograph by Stoleru Vasile).

Natural enemies (predators and parasites). This category includes methods in order to attract animals that eat insects and other harmful living animals.

Name of product	Pest	Crop	Pheromone/attractant	Application
Codling Moth Pagoda Trap with lure®	Codling moth	Apples	Pheromone attracts male moths, for monitoring only	8 traps/ha
Rollertrap®	Range of insects	Various	Two sided sticky trap	Yellow and blue
Pheromone trap®	Butterflies and moths	Protected and field crops	Monitors butterfly and moth population	5–8 traps/ha
Agralan Envirofleece®	Various pests	Protected and field crops	Polypropylene fleece, physical barrier to pests	17–30 g/m^2

Table 4. Pheromones and attractants for organic farming.

The effect of control pest in OA is to increase functional biodiversity, that is, to use wild flowers to attract parasitoids into the cabbage field—or to retain them if we release them—to increase natural pest control, directly through the added plants and the organisms that use them as resources and indirectly through the reduction of pesticides.

Creating proper shelters and feed for the useful fauna (frogs, green lizards, snakes, insectivore insects, and mammals), including their artificial breeding, have positive effects for farmers. Snakes can be used against rodents; hedgehogs counteract the attack of shell-less snails, mice, mole crickets, and also the Colorado beetles [39].

Predators catch and eat their prey. Some common predatory arthropods include ladybird beetles, carabid (ground) beetles, staphylinid (rove) beetles, syrphid (hover) flies, lacewings, minute pirate bugs, nabid bugs, big-eyed bugs, and spiders.

Entomophagy predators are species of animals which consume other animals, pests in particular.

The main species of insects and nematodes used for fighting against harmful insects are presented in **Table 5**. This method of biological control is widely used in horticulture, especially in protected areas, such as flower, orchard, and vegetables crops (**Figures 12–17**).

Name of products	Pests controlled	Crops	Parasites/predators	Application/dose
Aphipar®	Aphids (cotton aphid, peach and potato aphid, tobacco aphid	Protected crops	*Aphidius colemani* parasitic wasp	Preventive = 0.25 ex./m^2, curative light = 1 ex./m^2, curative heavy = 2 ex./m^2 (7 days interval, 3–6 application/year)
Ervipar®	Aphids (potato aphid, glasshouse potato aphid)	Protected crops	*Aphidius colemani* parasitic wasp	Preventive = 0.25 ex./m^2, curative light = 0.5 ex./m^2, curative heavy =2 ex./m^2 (7 days interval, continuously application)
Aphidend®	Aphids	Protected crops	*Aphidoletes aphidimyza* (gall midge)	Curative light = 1 ex./m^2, curative heavy = 10 ex./m^2 (continuously application)

Name of products	Pests controlled	Crops	Parasites/predators	Application/dose
Chrysoperla®	Aphids, whitefly, various thrips, caterpillars	Various	*Chrysoperla carnea* lacewing	Preventive = 0.5 ex./m², curative light = 1 ex./m², curative heavy = 5 ex./m² (continuously application)
Aphilin®	Aphids	Protected crops	*Aphelinus abdominalis* parasitic wasp	0.1–0.5 adult/m² for preventive use
Ervibank®	Aphids	Protected crops	*Aphidius ervi* parasitic wasp	Preventive = 0.5 ex./m² on interval of 14 days
Tricho-strip®	Caterpillars (lepidopteran eggs)	Protected crops	*Trichogramma brassicae* parasitic wasp	Preventive: min. 8 × 5 ex./m² each 7 days, curative light: min. 8 × 10 ex./m² each 7 days, curative heavy: min. 8 × 20/m² each 7 days
Fightacat®	Caterpillars (lepidopteran eggs)	Various	*Trichogramma evanescens* parasitic wasp	Preventive: min. 8 × 5 ex./m² each 7 days, curative light: min. 8 × 10 ex./m² each 7 days, curative heavy: min. 8 × 20/m² each 7 days
Anagrus®	Leafhopper	Protected	*Anagrus atomus* parasitic wasp	Preventive = 0.1 ex./m², curative heavy = 0.5 ex./m² (7 days interval)
Minex® Fightamine® Minusa®	Leaf miners	Protected crops	*Dacnusa sibirica* and *Diglyphus isaea* parasitic wasps	Preventive = 0,25 ex./m², curative light = 0.5 ex./m², curative heavy = 2 ex./m² (continuously application)
Miglyphus®	Leaf miners	Protected crops	*Diglyphus isaea* parasitic wasps	Preventive = 0.1 ex./m², curative heavy = 1 ex./m² (continuously application)
Cryptolaemus®	Mealy bug	Protected crops	*Cryptolaemus* sp. Australian ladybird	*Greenhouses:* 5 beetles per infested plant, *outdoors:* 1250–12,500 beetles per hectare *orchards:* 2500–5000 beetles per hectare
Spidex® Spidex-T® Fightamite A®	Mites (two spotted spider mite and carmine spider mite	Protected crops	*Phytoseiulus persimilis* predatory mite	Curative light = 0.5 ex./m², curative heavy = 2 ex./m² (continuously application every week)
Fightamite B®	Mites (two spotted spider mite)	Protected crops	*Feltiella acarisuga* (*Therodiplosis persicae*) predatory midge	250 adults/1000 m²
Typhlodromus®	Mites (red spider mite, two spotted spider mite, strawberry mite, broad mite, fruit tree spider mite)	Protected and outdoor crops	*Typhlodromus pyri* predatory mite	250–500 adults/1000 m²
Fightascale®	Soft scale insect	Protected crops	*Metaphycus helvolus* parasitic wasp	1 adult/m²
Entonem® Nemasys®	Sciarid flies	Various crops,	*Steinernema feltiae* nematode	200–400 ex./m²

Name of products	Pests controlled	Crops	Parasites/predators	Application/dose
Entomite®	Soil-living insects, thrips, collembola, nematodes, sciarid files	Various	*Hypoaspis aculeifer* or *Hypoaspis miles* predatory mites	Preventive = 100 ex./m^2, curative light = 200 ex./m^2, curative heavy = 500 ex./m^2 (one application)
Nemaslug® Slugsure®	Slugs	Various	*Phasmarhabditis hermaphrodita* nematode	500–1000 ex./m^2
Thripex® Fightathrip®	Thrips (various), spider mites	Protected crops	*Amblyseius cucumeris* predatory mite	Preventive = 50 ex./m^2, curative light = 100 ex./m^2(application at 14 days), curative heavy = 100 ex./m^2 (one application/week)
Thripor® Fightabug®	Thrips (various)	Protected crops	*Orius laevigatus*, *Orius insidiosus* or *Orius majusculus* predatory bug	Preventive = 0.5 ex./m^2, curative light = 1 ex./m^2, curative heavy = 10 ex./m^2 (one application/14 days)
Larvanem® Nemasys H®	Vine weevil	Various	*Heterorhabditis megidis* nematode	Curative light = 500,000/m^2, curative heavy = 1,000,000/m^2 (one application)
En-strip®	Whitefly	Protected crops	*Encarsia formosa* parasitic wasp	Preventive = 1.5–3 ex./m^2, curative light = 3–6 ex./m^2, curative heavy = 9 ex./m^2 (one applic./week)
Fightafly B®	Whitefly, leafhopper, leaf miner, spider mite	Protected crops	*Macrolophus caliginosus* predatory bug	Curative light = 10 ex./m^2, curative heavy = 50 ex./m^2 (one applic./14 days)

Table 5. Parasites and predators permitted for organic pest control.

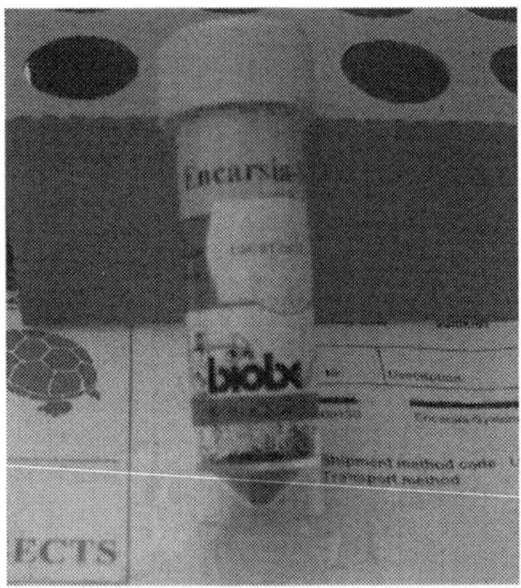

Figure 12. *Encarsia formosa* for greenhouse crops (photograph by Stoleru Vasile).

Figure 13. Applying the parasite wasp to a cucumber crop (photograph by Stoleru Vasile).

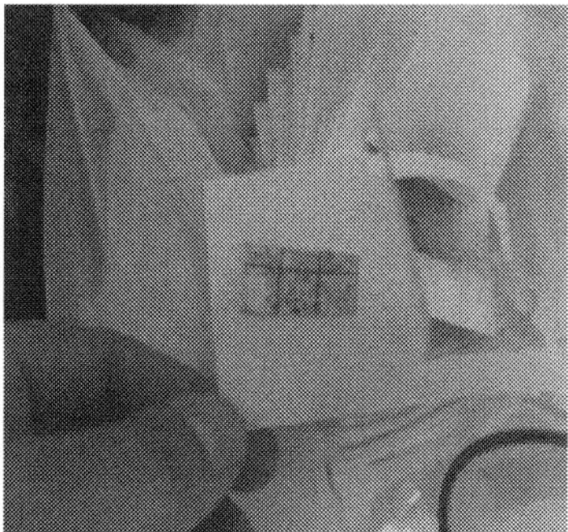

Figure 14. *Trichogramma* eggs plaques made in a laboratory (photograph by Stoleru Carmen).

Figure 15. Application of *Trichogramma* plaques for white butterfly eggs (photograph by Stoleru Carmen).

Figure 16. Tomato fruit damage by *Helicoverpa armigera* (photo by Deleanu Florina).

Figure 17. Whitefly in on the flower crop (photograph by Stoleru Vasile).

Biological methods. Biological control consists of using organisms and products against other living beings. The methods correspond to the future approaches; they are characterized by high selectivity and improbability levels regarding the fact of inducing the pest resistance phenomena, as well as a good capacity of self-perpetuation.

http://dx.doi.org/10.5772/64457

Economically speaking, these methods are more expensive, at least initially, when they have to be projected and produced, or when special installations are necessary and they require a lot of manual work for operation or for the uphill works. But in the end, does not the environment's health and ours implicitly deserve a bonus from the beneficiary?

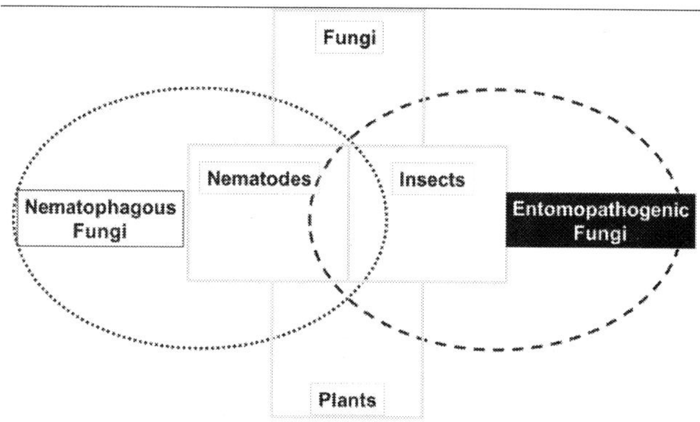

Figure 18. Multitrophic lifestyle of fungal parasites [38].

Figure 19. *B. bassiana* parasitism for *Bemisia tabaci* control (photograph by Sellitto Michele).

Microbiological control is a modern, efficient method but still quite expensive; it consists of using certain preparations based on living organisms (viruses, bacteria, fungi) that parasites and kill some of the pests.

Nowadays, more than 500 species of insect parasite fungi are known. Their advantage is that they spread out easily through spores and they are resistant to unfriendly conditions for long periods of time (**Figure 18**). In general, the relation between pests and their parasites are affected by global change, abiotic and biotic stresses to crops [40].

Beauveria sp. and *Metarhizium* sp. are two pathogenic fungi for insects which can penetrate the host insect through its exoskeleton due to its production of chitinolytic enzymes (**Figures 19** and **20**). Once inside the host, the fungus develops and feeds, causing its host's death.

The infested insects, still living, experience limited motion ability and the incapacity to feed themselves; moreover, they represent a source of infection for other insects [37].

Figure 20. *B. bassiana* on palm carbide (*Rhynchophorus ferrugineus*) (photograph by Sellitto Michele).

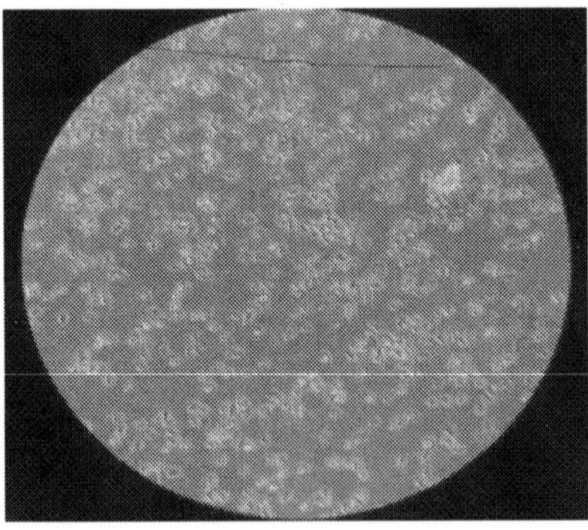

Figure 21. Conidi of *Pochonia chlamydospores* (photograph by Sellitto Michele).

Different studies have shown that *Beauveria* sp. and *Metarhizium* sp. actively control species from the following genera Coleoptera (*Melolontha* sp., *Diabrotica* sp.), Lepidoptera (*Tuta absoluta*), or Orthoptera [aphids, greenhouse whitefly, thrips, [41, 42] etc].

Pochonia sp. is a hyphomycete that acts as a parasite of nematode eggs. Its antagonistic activity is related to the production of proteolytic and chitinolytic enzymes that degrade the cellular structure of nematodes, especially that of eggs and females in the early stage (**Figures 21** and **22**).

Figure 22. Tubers of a potato attacked by nematodes (photograph by Aurelio Ciancio).

Arthrobotrys sp. is a fungus that parasitizes nematodes. The nematodes' biocontrol activity is related to the production of ring-like structures which swallow when a nematode pass by and catches it. Afterwards, the nematode is degraded by enzymes and used by the fungus as feed.

The combination between *Pochonia* sp. and *Arthrobotrys* sp. represents the most effective biological control method for the nematodes from a genera *Meloidogyne* sp., *Globodera* sp., and *Heterodera* sp. (**Figures 23** and **24**).

Figure 23. Adult of a nematode parasited by *Pochonia chlamydospores* (photograph by V.M. Sellitto, 2014).

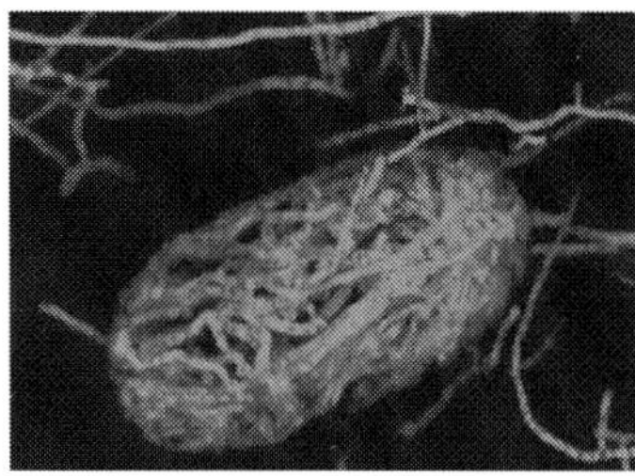

Figure 24. Egg of a nematode parasited by *Pochonia chlamydospores* (photograph by L. Lopez-Llorca, 2015).

The literature dealing with this subject mentions tests that proved that the use of these fungi, on soils sterilized using chemical products and solarization and steam, has maintained the soil and the level of nematodes below the damaging threshold for many years, compared to the soils where these fungi were not present [43].

Lecanicillium lecanii is a pathogenic fungus for numerous species of insects. This fungus acts as follows: the fungus spores lie and remain on the insects' exoskeleton, and then, they germinate and mechanically penetrate the insects' exoskeleton, due to their production of chitinolytic enzymes. From the industrial products containing entomopathogenic fungi, we mention the following: Muscardin M 45® and *Beauveria* spores (from *B. bassiana*), Boverin® (from *B. densa*), and Mitecidin® (from *Streptomyces aureus*), which act against the Colorado beetle and other coleopters (**Table 6**). Applying myco-insecticides, Naturalis-L® (*Beauveria bassiana*) and PreFeRal®WG (*Paecilomyces fumosoroseus*), were applied against adult *Rhagoletis cerasi* (Diptera: Tephritidae). In the first case, *B. bassiana* significantly reduced the number of damaged fruit (efficacy: 69–74%), whereas damage was not significantly reduced with PreFeRal®WG (efficacy: 27%) [44].

Name of product	Pest	Crop	Microorganisms	Dose/application
Vertalec®	Aphids	Protected crops	*Verticillium lecanii* fungal spores	2 g/L
Mycotal®	Whitefly, thrip larvae	Protected crops	*Verticillium lecanii* fungal spores	curative light = 0.1% (2–3 applic.), curative heavy = 0.1% (3–4 applic.)
Novosol FC® Dipel WP® Bactura WP®	Caterpillars	Vegetables, fruit, ornamentals	*Bacillus thuringiensis* var. kurstaki	1–1.6 kg/ha, depending of crop
Thuricide®	Caterpillars	Various	*Bacillus thuringiensis* var. *kurstaki*	2–4 tsp/gal water
Bactospeine®	Caterpillars	Various	*Bacillus thuringiensis* wettable powder	1–1.6 kg/ha, depending of crop

Table 6. Biological control agents used in organic farming.

Once the fungus is in, it develops and digests the insect from the inside until it kills it. Infested insects die in 4–6 days and are then covered with a whitish efflorescence, depending on the fungus sporulation. Thus, these insects become a source of infection for other insects. In addition, *Lecanicillium lecanii* can colonize certain tissues of the host plant, achieving an induced systemic resistance.

Many studies have shown that *L. lecanii* controls aphids, whitefly, and Thripidae genus. Other studies have proven that *Lecanicillium* sp. also controls certain nematode species as well as certain plant diseases, such as the gray mold (**Table 7**).

No.	Commercial name	Biological composition	Dose
1.	Pochar Linia Greenpower®	*Glomus* sp., *Pochonia* sp., *Arthrobotrys* sp.	2–3 l/ha in time and after transplantation
2.	Lecan Linia Greenpower®	*Glomus* sp., *Lecanicillium* sp.	2 l/ha or 0.2% foliar
3.	Metab Linia Greenpower®	*Glomus* sp., *Metarhizium* sp., *Beauveria* sp.	2 l/ha at the root or 0.2% foliar

Table 7. Microbiological products to control pests from vegetable crops.

From the bacteria used to fight against insects, *Bacillus thuringiensis* (**Figure 25**) and *B. subtilis* are the most popular (**Figure 26**).

Figure 25. *Bacillus thuringiensis var. kurstaki (photograph by V.M. Sellitto).*

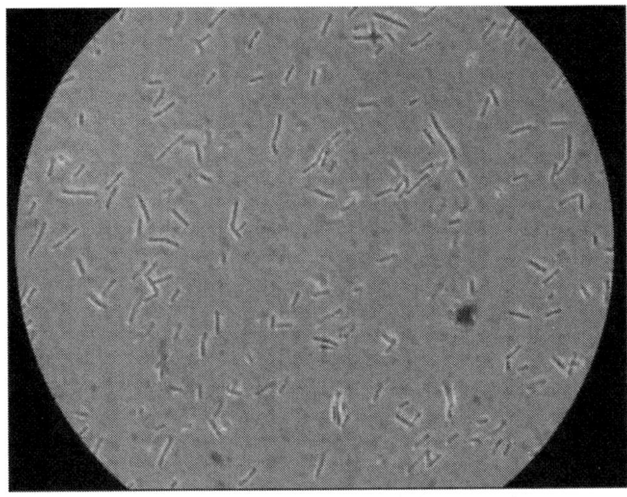

Figure 26. *Bacillus subtillis* (photograph by V.M. Sellitto).

During the last years, strains of *B. thuringiensis* were studied for their effect on the insect, through different toxins (**Table 8**).

No.	Toxins	Activities
1	Cry toxins	Pore formation on cell membrane; cytolysis activity
2	Vip toxins	Wide spectrum of insect activity
3	Thuricin	Bacteriocin
4	Hemolysin	Lysis of vertebrate red blood cells
5	Beta-exotoxins	Inhibition of RNA polymerase
6	Phospholipase-C	Cell membrane alteration

Table 8. Toxins produced by strains of *Bacillus thuringiensis*.

It laid at the basis of the process of obtaining numerous commercial products: Agritol®, Dipel®, Thuricide®, Novodor 3FC®, Vectobac®, Bactospeine®, Thuringine®, Entobakterin®, Thurintox®, or Foray®. These products are highly efficient in counteracting the larvae of certain butterflies from vegetables crops [37].

Out of more than 300 viruses that cause diseases for more than 175 species of insects, polyhedric viruses are the most known; they are used at obtaining certain preparations industrially, such as Biotrol VHZ® and VSE®, Vitex® (against caterpillars), and Virin-ENS® (recommended in fighting against the cabbage moth). Nuclear polyhedrosis viruses (NPV) and granulosis viruses (GV) are available to get rid of some caterpillar pests (*Mamestra brassicae, Helicoverpa armigera, Autographa gamma, Pieris brassicae, and Euproctis chrysorrhoea*) [45] (**Figure 27**).

Figure 27. Uninfected (bottom) beet armyworm (*Spodoptera exigua*) and beet armyworm killed by the nuclear polyhedrosis virus. Photograph credit: David Nanace, USDA ARS.

Genetic methods. The works of ameliorating plants have as their main objective the production of cultivars endowed with greater resistance. This is why the forms providing higher mechanical resistance are promoted (with thicker cuticle or suber, with a waxy protective layer or with abundant porosity), physiological or chemical (by growing the content of substances with repellent or insecticide effect).

Several aphid species can proliferate in winter lettuce crops, such as *Nasonovia ribisnigri* (Mosley), *Myzus persicae* (Sul.), *Aulacorthum solani* (Kalt.), *Macrosiphum euphorbiae* (Th.), and *Hyperomyzus lactucae* (L.). *N. ribisnigri* is the most damaging one because it preferentially develops in the lettuce heart [46, 47]. In addition to feeding damage and the loss of product quality due to their presence when the lettuce is marketed, aphids are also vectors of viruses, such as the lettuce mosaic virus. Finally, slugs (*Deroceras* sp. and *Arion* sp.) and snails can also cause feeding damage to lettuce in winter.

Complete resistance to the aphid *N. ribisnigri* and partial resistance to *M. persicae* are conferred by a dominant gene called Nr, which has been introduced in many European cultivars [48]. However, this resistance was recently bypassed by a new *N. ribisnigri* biotype named Nr:1 [49].

2.2.3. *Using plants to fight against pests*

This method relies on certain plants' feature of secreting in the earth or in the air certain substances with repulsive or destructive effects on pests. By and large, these plants can be cultivated in the field, as border or associated with the crops. The important species with insecticide effect are presented in **Table 9**.

Biochemical methods. The products used for protecting plants against harmful insects can be classified according to the raw material used, into two categories: vegetal insecticides and mineral insecticides.

Species	Controlled pests
Yarrow (*Achillea millefolium*)	Aphids, mites, psyllids, thrips
Queen of poisons (*Aconitum* sp.)	Coleopteran larvae
Sweet flag (*Acorum calamus*)	White cabbage butterfly
Onion (*Allium cepa*)	Mites, ants, storehouse pests
Garlic (*Allium sativum*)	Thrips, storehouse pests
Birthwort (*Aristolochia clematitis*)	Bed bug
Absinthium (*Artemisia absinthium*)	Nematodes, caterpillars, fleas
Mugwort (*Artemisia vulgaris*)	Fleas, Colorado beetle
Lamb's quarters (*Chenopodium album*)	Colorado beetle, white butterfly
Hemlock (*Conium maculatum*)	Coleopteran larvae
Coriander (*Coriandrum sativum*)	Aphids, spiders, Colorado beetle (repellent effect)
Spurge (*Euphorbia* sp.)	Caterpillars, aphids
White sweet clover (*Melilotus albus*)	Colorado beetle
Mint (*Mentha* sp.)	Colorado beetle
Tobacco (*Nicotiana tabacum*)	Aphids, mites, Colorado beetle
Black nightshade (*Solanum nigrum*)	Aphids, mites, Colorado beetle, cabbage butterfly
Yew (*Taxus baccata*)	Various insects
Field penny-cress (*Thlaspi arvense*)	Bed bug (repellent)
Common nettle (*Urtica dioica*)	Aphids, mites
Mullein (*Verbascum phlomoides*)	Colorado beetle

Table 9. Plants used in organic farming with a repellent effect.

Vegetal insecticides. Insecticides of natural origin are substances which can cause the death of insects interfere in the development or reproduction being responsible to attract or repel them. Today, worldwide, there are more than 1450 species of plants with insecticide effects, from which only approximately 50 are useful [1]. As far as our country is concerned, too little from the 200 species credited with this action have been or are being effectively used in this purpose, and even fewer have been studied from this point of view.

Stinging nettle (*Urtica dioica*). Action: it stimulates plant growth, it slows down the attack of certain insects, counteracts aphids, and spiders before the formation of leaves and flowers [37].

Fern (*Dryopteris filix-mas*). Leaf purine and decoction, undiluted, are used against shell-less snails (every time needed). At the same time, this product, diluted 10 times with water, is used for the late spring treatments against aphids.

Wormwood (*Artemisia absinthium*). This plant can be used as an undiluted purine (caterpillars, lice), cold extract diluted twice for Solanaceae against the larvae of the Colorado beetle [37], or decoction is used undiluted against the cabbage fly [2].

Tansy (*Tanacetum vulgare*) is used as an undiluted infusion every time it is needed against ants, aphids, fleas and other insects.

Wild garlic (*Allium ursinum*). Wild garlic infusion is used undiluted, by repeatedly aspersing the plants every 3 days against aphids and mites. Purine is also used undiluted against the carrot fly (*Psila rosae*), but only during its flight period.

Garlic (*Allium sativum*). It can be used in the treatment of mites and also in seed treatments. Garlic in its natural state is eventually cultivated in rows, has a nematode effect (*Meloidogyne* sp.), and drives away the striped field mouse.

Scientific name	Common name	Scientific name	Common name
Leptinotarsa decemlineata	Colorado beetle	*Acyrthosiphon pisum*	Pea aphid
Mamestra brassicae	Cabbage moth	*Pieris brassicae*	Large white
Pieris rapae	Small white	*Trialeurodes vaporariorum*	Greenhouse whitefly
Gnorimoschema lycopersicella	Tomato pinworm		

Table 10. The action spectrum of the *Chrysanthemum cinerariaefolium* extract.

Pyrethrum (Chrysanthemum cinerariaefolium, Pyrethrum cinerariaefolium). Pyrethrum is a contact insecticide having paralyzing effect and a wide range of actions. The great advantage, ecologically speaking, is that it completely decomposes into harmless compounds in only 48 h after application [50]. Pyrethrum is noticed on a large number of insects and mites with a soft body or when they are still in a larval stage, as a solution with concentration of 0.1% (**Table 10**). The extract of pyrethrum cannot recommend mixture with alkaline products, Bordeaux mixture [1, 39].

Derris powder (*Derris* sp.). Derris powder is applied to a large number of aphids, nematodes, and insects, more vulnerable as their ingestion capacity is larger (larvae). Its toxicity for warm blooded animals is null, while for the other ones, it is lethal, used as decoct of ground fresh or dried roots, in a solution of 0.01%.

Gliricidia (*Gliricidia sepium*). Action: repellent, parasitic, rodenticide, mixed with grain, left from place to place on a field or put in warehouses; in a few days, it kills the rodents [9, 51].

Neem (*Azadirachta indica*). It is a repellent, hormonal disruptive (it blocks the larval metamorphosis process), nematocide and antimicotic. Azadirachtin is extracted from this plant's seeds, the active substance of NeemAzal T/S®.

The preparations destroy the eggs, larvae, and adults of more than 200 species of field or storehouse pests in the case of beans, cereals, tomatoes, and field plants from the most various classes: nematodes, ants, bed bugs, grasshoppers, etc. Neem oil is used in fighting against certain pests on plants, and ground marc has a nematode effect [33].

Bitter wood (*Quassia amara*). The active substances of this preparation act as contact and ingestion insecticide but are slower than pyrethrum. It is used in fighting against many pests: aphids, flies, cabbage aphids, etc.

Decoct is made from 100 to 150 g chips of bitter wood at 10 l water. The bitter wood decoction can be improved by adding an equal amount of solution of potassium soap in a concentration of 1–2.5% [51].

Traditionally, in organic fruit growing, the apple sawfly *Hoplocampa testudinea* Klug is controlled by the use of extracts of bitter wood of 6 g/ha/in 500 l. For a good efficiency, the bitter extract can be mixed with Nemmazal T/S® [52, 53].

Name of product	Pests	Crops	Agent for control	Dose/concentration
Savona®	Whiteflies, thrips, aphids, mealy bugs, leafhoppers	Various	Fatty acid	1–2%, one applic./week
Liquid Derris®	Various biting and sucking insects	Various	Rotenone derived from *Lonchocarpus utilis and L. urucu*	0.8–1 l/ha
Bug-Me-Not Bloom and Leaf Astringent Spray (CP) ®	Various insects	Various	Insect repellent based on neem extract	Insect repellent, 4–6 tsp./10 l
Bug-Me-Not Root and Soil Granules (CP) ®	Various insects	Various	Insect repellent based on neem extract	Insect repellent, 4–6 tsp./10 l
AquaPy®	For insects in grain stores	Store	Natural pyrethrum. organic products (e.g., grain) must be removed before use	1 l/3000 m^3
Jet 5®	For cleaning glasshouses /polytunnels	Store	Peroxyacetic acid	0.2%

Table 11. Commercial products permitted to use in organic farming.

2.2.4. *Repellent mineral products*

Potassium alum. This preparation is used as solution with a concentration of 0.4% with good efficacy against lice and caterpillars. At the same time, aspersing the soil with this solution is quite efficacious against shell-less snails. Basalt flour. It is used as a powder. Its action against pests is explained because of a change of the pH at the surface of aerial organs from weak acid (preferred by most pests) to weak alkaline or mechanical action on the insect's body, their eyes, and trachea [2].

2.2.5. Insecticide mineral preparations

Potassium soap is successfully used against mites (red spider) and the cabbage aphid. The treatment is applied alone or in combination with other products (horsetail extract) by repeatedly aspersing the plants with various solution types: 200–300 g soap at 10 l water (lice); 200–300 g soap + 0.5 l alimentary alcohol + 1 table-spoonful of lime and 1 table-spoonful of cooking salt at 10 l of water, against the red spider and the larvae of the Colorado beetle [39].

The preparation is used as solution with concentration of 1–2% with good efficacy against lice and leaf fleas, found under the name Neudosan® or Savona® [9], like as other products presented in **Table 11**.

Acknowledgements

The authors wish to thank Mr. Dean Hufstetler for reviewing article.

Author details

Vasile Stoleru[1*] and Vicenzo Michele Sellitto[2]

*Address all correspondence to: vstoleru@uaiasi.ro

1 Department of Horticulture, University of Agricultural Sciences and Veterinary Medicine, Iasi, Romania

2 Department of Applied Microbiology, Microspore Ltd., Larino, Italy

References

[1] Toncea I. Practical Guide for organic agriculture. Organic technologies for land cultivation. Cluj-Napoca: Publisher Academic Press 2002.

[2] Stoleru V, Munteanu N, Sellitto VM. New approach of organic vegetable systems. Aracne Editrice; Rome 2014.

[3] EC Regulation 834. Organic production and labelling of organic products and repealing Regulation (EEC) No 2092/91, (2007).

[4] EC Regulation 889. Laying down detailed rules for the implementation of Council Regulation (EC) No 834/2007 on organic production and labelling of organic products with regard to organic production, labelling and control, (2008).

[5] Willer H. The World of Organic Agriculture. Statistics and Emerging. Trends 2015. Frick: FIBL; 2015. 306 p.

[6] Knipling EF. Entomology and the Management of Man's Environment. Australian Journal of Entomology. 1972;11(3):153–67.

[7] Hatman M, Bobes I, Lazar A, Gheorghies C, Glodeanu C, Severin V, Tusa C, Popescu I, Vonica I. Phytopatology. Bucharest: Editura Didactica si Pedagogica; 1989. 468 p.

[8] Georgescu T. Horticultural entomology: Dosoftei Publisher; 2006. 426 p.

[9] Stoleru V, Albert IO. Vegetable growing using organic measures. Risoprint; Cluj-Napoca 2007.

[10] Yun S-H, Koo H-N, Kim HK, Yang J-O, Kim G-H. X-ray irradiation as a quarantine treatment for the control of six insect pests in cut flower boxes. Journal of Asia-Pacific Entomology. 2016;19(1):31–8.

[11] Osouli S, Ziaie F, Haddad Irani Nejad K, Moghaddam M. Application of gamma irradiation on eggs, active and quiescence stages of *Tetranychus urticae* Koch as a quarantine treatment of cut flowers. Radiation Physics and Chemistry. 2013;90:111–9.

[12] Boiteau G, Heikkilä J. Chapter 12 - Successional and Invasive Colonization of the Potato Crop by the Colorado Potato Beetle: Managing Spread. In: Alyokhin A, Giordanengo CV, editors. Insect Pests of Potato. San Diego: Academic Press; 2013. p. 339–71.

[13] Zeleke KT, Nendel C. Analysis of options for increasing wheat (*Triticum aestivum* L.) yield in south-eastern Australia: The role of irrigation, cultivar choice and time of sowing. Agricultural Water Management. 2016;166:139–48.

[14] Pfiffner L, Wyss E. Use of sown wildflower strips to enhance natural enemies of agricultural pests. In: Gurr GM, Wratten SD, Altieri MA, editor. Ecological Engineering for Pest Management: Advances in Habitat Manipulation for Arthropods. Collingwood, VIC, Australia: CSIRO Publishing; 2004. p. 167–88.

[15] Munteanu N. Tomatoes, peppers and eggplants. Iasi: Ion Ionescu de la Brad; 2003. 214 p.

[16] Indrea D, Apahidean S, Apahidean M, Maniutiu D, Sima R. Vegetable growing Editor. Bucuresti: Ceres; 2012. 624 p.

[17] Ciofu R, Stan N, Popescu V, Chilom P, Apahidean S, Horgos A, Berar V, Lauer KF, Atanasiu N. Tratat de legumicultura. Bucuresti: Editura Ceres; 2004. 1165 p.

[18] López-Pérez J-A, Le Strange M, Kaloshian I, Ploeg AT. Differential response of Mi gene-resistant tomato rootstocks to root-knot nematodes (*Meloidogyne incognita*). Crop Protection. 2006;25(4):382–8.

[19] Schneider JHM, s'Jacob JJ, van de Pol PA. Rosa multiflora 'Ludiek', a rootstock with resistant features to the root lesion nematode *Pratylenchus vulnus*. Scientia Horticulturae. 1995;63(1–2):37–45.

[20] King SR, Davis AR, Zhang X, Crosby K. Genetics, breeding and selection of rootstocks for Solanaceae and Cucurbitaceae. Scientia Horticulturae. 2010;127(2):106–11.

[21] Stoleru V, Munteanu N, Stoleru CM, Rotaru L. Cultivar selection and pest control techniques on organic white cabbage yield. Notulae Botanicae Horti Agrobotanici. 2012;40(2):190–6.

[22] Vilela M, Moraes JC, Coelho M, Françoso J, Santos-Cividanes TMd, Sakomura R. Response of moderate pest resistant and susceptible cultivar of sugarcane to silicon application. American Journal of Plant Sciences. 2014;5(26):3823–8.

[23] Abdel-Sabour AG, Obiadalla-Ali HA, AbdelRehim KA. Genetic and chemical analyses of six cowpea and two Phaseolus bean species differing in resistance to weevil pest. Journal of Crop Science and Biotechnology. 2010;13(1):53–60.

[24] López-Gómez M, Flor-Peregrín E, Talavera M, Verdejo-Lucas S. Suitability of zucchini and cucumber genotypes to populations of *Meloidogyne arenaria*, *M. incognita*, and *M. javanica*. Journal of Nematology. 2015;47(1):79–85.

[25] Andersson SC, Johansson E, Baum M, Rihawi F, El-Bouhssini M. New resistance sources to Russian wheat aphid (*Diuraphis noxia*) in Swedish wheat substitution and translocation lines with rye (Secale cereale) and Leymus mollis. Czech Journal of Genetics and Plant Breeding. 2015;51(4):162–5.

[26] Silva SZd, Alves LFA, Coelho SRM, Tessaro D. Technological and nutritional quality of cowpea (*Vigna unguiculata* (L.) Walp.) cultivars infested by *Callosobruchus maculatus* (Fabr.) (Coleoptera: Bruchidae). Journal of Food, Agriculture and Environment. 2014;12(2):731–4.

[27] Maldonado Moreno N, Ascencio Luciano G, Gill Langarica HR. Huasteca 300, a new soybean cultivar for the south of Tamaulipas state. Agricultura Técnica en México. 2009;35(4):481–5.

[28] Lopez-Llorca LV, Olivares-Bernabeu C, Salinas J, Jansson H-B, Kolattukudy PE. Pre-penetration events in fungal parasitism of nematode eggs. Mycological Research. 2002;106(4):499–506.

[29] Ciancio A, Bonsignore R, Vovlas N, Lamberti F. Host records and spore morphometrics of *Pasteuria penetrans* group parasites of nematodes. Journal of Invertebrate Pathology. 1994;63(3):260–7.

[30] Sabareanu-Stoleru C-M. Research on the technology for white cabbage growing (*Brassica oleracea* L., var. *capitata* L., f. *alba* DC) in ecological vegetable system, in conditions of Iasi county [Doctoral thesis]. Iasi: University of Agricultural Sciences and Veterinary Medicine; 2010.

[31] Gill HK, McSorley R. Impact of different organic mulches on the soil surface arthropod community and weeds. International Journal of Pest Management. 2012;58(1):33–40.

[32] Gill HK, McSorley R, Treadwell DD. Comparative performance of different plastic films for soil solarization and weed suppression. HortTechnology. 2009;19(4):769–774.

[33] Brian Caldwell ES, Abby S, Anthony S, Christine S. Resource Guide for Organic Insect and Disease Management. Ithaca, New York: Arnold Printing Corp; 2013. 201 p.

[34] Linker HM, Orr DB, Barbercheck ME. Insect management on organic farms. Center for Environmental Farming Systems. 2009:1–37.

[35] Stan N, Munteanu N, Stan T. Vegetable Growing. Iasi: Ion Ionescu de la Brad 2003. 316 p.

[36] Lampkin N. The Principles of Organic Farming. U.K.: Farming Press, Miller Freeman; 2001.

[37] Calin M. The guide of the recognition and pest control of vegetable in biological agriculture. Bacau: Tipoactiv; 2005. 376 p.

[38] Lind K, Lafer G, Schloffer K, Innerhofer G, Meister H. Organic Fruit Growing. Wallingfors: CABI Publishing; 2003. 282 p.

[39] Stoleru V. The management of organic vegetable systems. Iasi: Ion Ionescu de la Brad; 2013. 250 p.

[40] Lopez-Llorca L, editor Fungal parasites of invertebrates: useful tools for adapting crops to global change. International Congress Soil and Food, Resources for a Healthy Life; 2015; Iasi: UASVM Iasi.

[41] Pires L, Marques E, Wanderley-Teixeira V, Teixeira A, Alves L, Alves S. Ultrastrucure of *Tuta absoluta* parasitized eggs and the reproductive potential of females after parasitism by *Metharhizium anisopliae*. Micron. 2009;40:255–261.

[42] Lecuona R, Riba G, Cassier P, Clement JL. Alterations of insect epicuticular hydrocarbons during infection with *Beauveria bassiana* or *B. brongniartii*. Journal of Invertebrate Pathology. 1991;58(1):10–18.

[43] McSorley R, Wang KH, Rosskopf EN, Kokalis-Burelle N, Hans Petersen HN, Gill HK, Krueger R. Nonfumigant alternatives to methyl bromide for management of nematodes, soil-borne diseases, and weeds in production of snapdragon (*Antirrhinum majus*). International Journal of Pest Management. 2009;55(4):265–273.

[44] Daniel C, Wyss E. Field applications of Beauveria bassiana to control the European cherry fruit fly *Rhagoletis cerasi*. Journal of Applied Entomology. 2010;134(9–10):675–81.

[45] Kalha CS, Singh PP, Kang SS, Hunjan MS, Gupta V, Sharma R. Entomopathogenic viruses and bacteria for insect-pest control. Integrated Pest Management. Academic Press, Elsevier 2014; pp. 225–244.

[46] Scorsetti AC, Maciá A, Steinkraus DC, López Lastra CC. Prevalence of Pandora neoaphidis (Zygomycetes: Entomophthorales) infecting *Nasonovia ribisnigri* (Hemiptera: Aphididae) on lettuce crops in Argentina. Biological Control. 2010;52(1):46–50.

[47] Liu YB. Distribution and population development of *Nasonovia ribisnigri* (homoptera: Aphididae) in iceberg lettuce. J Econ Entomol. 2004;97:883–90.

[48] Shrestha G, Enkegaard A, Steenberg T. Laboratory and semi-field evaluation of *Beauveria bassiana* (Ascomycota: Hypocreales) against the lettuce aphid, *Nasonovia ribisnigri* (Hemiptera: Aphididae). Biological Control. 2015;85:37–45.

[49] Ten Broeke CJM, Dicke M, Van Loon JJA. Feeding behaviour and performance of different populations of the black currant-lettuce aphid, *Nasonovia ribisnigri,* on resistant and susceptible lettuce. Entomologia Experimentalis et Applicata. 2013;148:130–41.

[50] Stan N, Munteanu N. Legumicultura. Iasi: Ion Ionescu de la Brad; 2001. 242 p.

[51] Munteanu N, Stoleru V, Filipov F, Teliban G, Lorena-Diana P. To assess the climatic potential of biological vegetable in Iasi county. Grant 31/2006, 2006 financed by UEFISCDI.

[52] Sjöberg P, Swiergiel W, Neupane D, Lennartsson E, Thierfelder T, Tasin M, Rämert B. Evaluation of temperature sum models and timing of *Quassia amara* (Simaroubaceae) wood-chip extract to control apple sawfly (*Hoplocampa testudinea* Klug) in Sweden. Journal of Pest Science. 2015;88(2):301–10.

[53] Kienzle J, Kopp B, Schulz C editor. Control of the apple sawfly (*Hoplocampa testudinea* Klug) with extracts from *Quassia amara* L.: Quality and combination. 10th International Conference on Cultivation Technique and Phytopathological Problems in Organic Fruit-Growing; 2002. Weinsberg, Germany: Proceedings to the Conference from 31st January.

Chapter 3

Entomopathogenic Nematodes in Pest Management

Ugur Gozel and Cigdem Gozel

Additional information is available at the end of the chapter

http://dx.doi.org/10.5772/63894

Abstract

The definition "biological control" has been used in different fields of biology, most notably entomology and plant pathology. It has been used to describe the use of live predatory insects, entomopathogenic nematodes (EPNs) or microbial pathogens to repress populations of various pest insects in entomology. EPNs are among one of the best biocontrol agents to control numerous economically important insect pests, successfully. Many surveys have been conducted all over the world to get EPNs that may have potential in management of economically important insect pests. The term "entomopathogenic" comes from the Greek word *entomon* means insect and *pathogenic* means causing disease and first occurred in the nematology terminology in reference to the bacterial symbionts of *Steinernema* and *Heterorhabditis*. EPNs differ from other parasitic or necromenic nematodes as their hosts are killed within a relatively short period of time due to their mutualistic association with bacteria. They have many advantages over chemical pesticides are in operator and end-user safety, absence of withholding periods, minimising the treated area by monitoring insect populations, minimal damage to natural enemies and lack of environmental pollution. Improvements in mass-production and formulation technology of EPNs, the discovery of numerous efficient isolates and the desirability of increasing pesticide usage have resulted in a surge of scientific and commercial interest in these biological control agents.

Keywords: biological control, safety, entomopathogenic nematodes, *Steinernema*, *Heterorhabditis*

1. Entomopathogenic nematodes

1.1. General information of entomopathogenic nematodes

Entomopathogenic nematodes (EPNs) are soil-inhabiting, lethal insect parasites that belong to the Phylum Nematoda from the families Steinernematidae and Heterorhabditidae, and they have proven to be the most effective as biological control organisms of soil and above-ground pests [1, 2]. They have been known since the seventeenth century [3], but it was only in the 1930s that serious care was given by using nematodes for pest control.

So far, the family Steinernematidae is comprised of two genera, *Steinernema* Travassos, 1927 [4] (Poinar, 1990) and *Neosteinernema* (Nguyen and Smart, 1994) [5]. *Neosteinernema* contains only one species *Neosteinernema longicurvicauda* that isolated from the termite *Reticulitermes flavipes* (Koller). The family Heterorhabditidae contains only one genus, *Heterorhabditis* Poinar, 1976 [6].

EPNs are mutually associated with bacteria of the family Enterobacteriaceae; the bacterium carried by Steinernematidae is usually a species of the genus *Xenorhabdus*, and that carried by Heterorhabditidae is a species of *Photorhabdus*. The third juvenile stage of EPNs is referred to as the "infective juvenile" (IJ) or the "dauer" stage. IJs of both genera release their bacterial symbionts in the insect host body and develop into fourth-stage juveniles and adults. The insects die mainly due to a septicemia. Sometimes a bacterial toxaemia precedes the resulting septicemia [7].

Infective juvenile is the only free-living stage and can survive in soil for several months until susceptible insects are encountered. IJs locate and infect suitable insect hosts by entering the insect host through the mouth, anus, spiracles or thin parts of the host cuticle. After infection, the symbiotic bacteria are released into the insect haemocoel, causing septicaemia and death of the insect [1, 8]. When an insect host is infected in the soil by an EPN, development and reproduction within the cadaver can take 1–3 weeks [9].

Surveys for EPNs have been conducted in temperate, subtropical and tropical regions and found that EPNs have a worldwide distribution; the only continent where they have not been found is Antarctica [10]. Soil texture, temperature and host availability are thought to be important factors in determining their distribution [11–13].

Nearly 70 valid species of *Steinernema* [14–16] and 25 species of *Heterorhabditis* [17, 18] have been described worldwide and still surveys for EPNs have been conducted in many parts of the world.

1.2. Biology and life cycle of entomopathogenic nematodes

Through all nematodes studied to control insects, the families Steinernematidae and Heterorhabditidae have made a sensation and information about them is increasing exponentially. Steinernematids and Heterorhabditids from these families have similar life cycles, and the only difference between the life cycles of *Heterorhabditis* and *Steinernema* is occurred in the first generation. *Steinernema* species are amphimictic; this means that for successful reproduction they require the presence of males and females, whereas *Heterorhabditis* species are hermaphroditic and able to reproduce in the absence of conspecifics.

Both nematode genera reproduction is amphimictic in the second generation [4]. However, a hermaphroditic Steinernematid species was isolated from Indonesia [19]. Only the free-living, IJ stage is able to target insect host and the only form found outside of the host. EPNs occur naturally in soil and locate their host in response to carbon dioxide, vibration and other chemical cues, and they react to chemical stimuli or sense the physical structure of insect's integument [1].

IJs penetrate the host insect via the spiracles, mouth, anus, or in some species through intersegmental membranes of the cuticle, and then enter into the haemocoel [20]. IJs release cells of their symbiotic bacteria from their intestines into the haemocoel. The bacteria multiply rapidly in the insect hemolymph, provide nematode with nutrition and prevent secondary invaders from contaminating the host cadaver, and the infected host usually dies within 24–48 hours by bacterial toxins.

Nematodes reproduce until the food supply becomes limiting at which time they turn into IJs. The progeny nematodes go through four juvenile stages to the adult. Based on the available resources, one or more generations may occur within the host cadaver, and a great number of IJs are released into environment to infect other host insects and continue their life [1].

The insect cadaver becomes red if the insects are killed by Heterorhabditids and brown or tan if killed by Steinernematids (**Figure 1**). The colour of the insect host body is indicative of the pigments produced by the monoculture of mutualistic bacteria growing in the host insects [1].

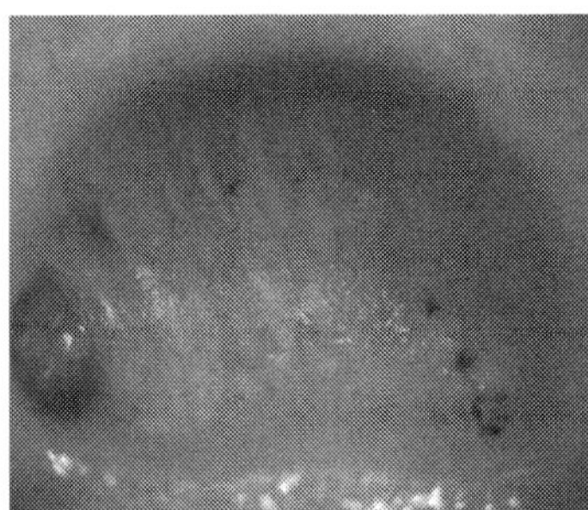

Figure 1. Different colours of the dead *Curculio nucum* larvae on white traps after EPNs infection.

The foraging strategies of EPNs change between species, and they use two main foraging strategies: ambushers or cruisers [21]. *Steinernema carpocapsae* is an example of ambushers, which have an energy-conserving approach and lie in wait to attack mobile insects (nictitating) in the upper layer of the soil. *Steinernema glaseri* and *Heterorhabditis bacteriophora* are examples of cruisers are highly active and generally subterranean, moving significant distances using volatile cues and other methods to find their host underground. But they are also successful to attack white grubs (Scarab beetles), which are less mobile. Other species, such as *Steinernema feltiae* and *Steinernema riobrave*, use an intermediate foraging strategy (combination of ambush and cruiser type) to find their host.

Selection of an EPN to control a particular pest insect is based on various factors: the nematode's host range, host finding or foraging strategy, tolerance of environmental factors and their effects on survival and efficacy. The most critical factors are moisture, temperature, pathogenicity for the targeted pest insect and foraging strategy [1, 22–24]. The activity, infectivity and survival of EPNs can be profoundly influenced by soil composition, through its effects on moisture retention, oxygen supply and texture [25–27].

Within a favourable range of temperatures, adequate moisture and a susceptible host, those EPNs with a mobile foraging strategy (cruisers and intermediate foraging strategies) could be considered for use in subterranean and certain above-ground habitats (foliar, epigeal and cryptic habitats). Those EPNs with a sit and wait foraging strategy (ambushers) will be most effective in cryptic and soil surface habitats [28].

1.3. Advantages of entomopathogenic nematodes

These nematodes have many advantages; EPNs and their associated bacterial symbionts have been proven safe to warm-blooded vertebrates, including humans [29, 30]. Cold-blooded species have been found to be susceptible to EPNs under experimental conditions at very high dosages [31, 32]. However, under field conditions, the negative results could not be reproduced [33, 34].

Most biological agents require days or weeks to kill the host, yet nematodes can kill insects usually in 24–48 hours. They are easy and relatively inexpensive to culture, live from several weeks up to months in the infective stage, are able to infect numerous insect species, occur in soil and have been recovered from all continents except Antarctica [1, 35].

Foliar applications of nematodes have been successfully used to control the quarantine leaf-eating caterpillars as *Tuta absoluta, Spodoptera littoralis, Helicoverpa armigera, Pieris brassicae* on several crops and have the potential for controlling various other insect pests. Application of EPNs does not require masks or other safety equipment like chemicals. EPNs and their associated bacteria have no detrimental effect to mammals or plants [29, 30, 36].

2. Use of entomopathogenic nematodes

Potential of EPNs as insecticidal agents has been tested against a wide range insect species by many researchers all over the world. They have been used with different success against insect pests occurred in different habitats. Much success has been obtained against soil-dwelling pests or pests in cryptic habitats such as inside galleries in plants where IJs find excellent atmosphere to survive and protect themselves from environmental factors. Commercial use of EPNs against some pest insects is given in **Table 1**.

Crops (targeted)	Pest common name	Pest scientific name	Effective nematodes[b]
Artichokes	Artichoke plume moth	*Platyptilia carduidactyla*	Sc
Vegetables	Armyworm	Lep: Noctuidae	Sc, Sf, Sr
Ornamentals	Banana moth	*Opogona sacchari*	Hb, Sc
Bananas	Banana root borer	*Cosmopolites sordidus*	Sc, Sf, Sg
Turf	Billbug	*Sphenophorus* spp. (Col: Curculionidae)	Hb, Sc
Turf, vegetables	Black cutworm	*Agrotis ipsilon*	Sc
Berries, ornamentals	Black vine weevil	*Otiorhynchus sulcatus*	Hb, Hd, Hm, Hmeg, Sc, Sg
Fruit trees, ornamentals	Borer	*Synanthedon* spp. and other sesiids	Hb, Sc, Sf
Home yard, turf	Cat flea	*Ctenocephalides felis*	Sc
Citrus, ornamentals	Citrus root weevil	*Pachnaeus* spp. (Col: Curculionidae)	Sr, Hb
Pome fruit	Codling moth	*Cydia pomonella*	Sc, Sf
Vegetables	Corn earworm	*Helicoverpa zea*	Sc, Sf, Sr
Vegetables	Corn rootworm	*Diabrotica* spp.	Hb, Sc
Cranberries	Cranberry girdler	*Chrysoteuchia topiaria*	Sc
Turf	Crane fly	Dip: Tipulidae	Sc
Citrus, ornamentals	Diaprepes root weevil	*Diaprepes abbreviatus*	Hb, Sr
Mushrooms	Fungus gnat	Dip: Sciaridae	Sf, Hb
Grapes	Grape root borer	*Vitacea polistiformis*	Hz, Hb
Iris	Iris borer	*Macronoctua onusta*	Hb, Sc
Forest plantings	Large pine weevil	*Hylobius abietis*	Hd, Sc
Vegetables, ornamentals	Leafminer	*Liriomyza* spp. (Dip: Agromyzidae)	Sc, Sf
Turf	Mole cricket	*Scapteriscus* spp.	Sc, Sr, Sscap
Nut and fruit trees	Navel orangeworm	*Amyelois transitella*	Sc
Fruit trees	Plum curculio	*Conotrachelus nenuphar*	Sr
Turf, ornamentals	Scarab grub[c]	Col: Scarabaeidae	Hb, Sc, Sg, Ss, Hz
Ornamentals	Shore fly	*Scatella* spp.	Sc, Sf
Berries strawberry	Root weevil	*Otiorhynchus ovatus*	Hm
Bee hives	Small hive beetle	*Aethina tumida*	Hi, Sr
Sweet potato	Sweetpotato weevil	*Cylas formicarius*	Hb, Sc, Sf

[a]Nematodes listed provided at least 75% suppression of these pests in field or greenhouse experiments.
[b]Abbreviations of nematode species; Hb: *Heterorhabditis bacteriophora*, Hd: *H. downesi*, Hi: *H. indica*, Hm: *H. marelata*, Hmeg: *H. megidis*, Hz: *H. zealandica*, Sc: *Steinernema carpocapsae*, Sf: *S. feltiae*, Sg: *S. glaseri*, Sk: *S. kushidai*, Sr: *S. riobrave*, Sscap: *S. scapterisci*, Ss: *S. scarabaei*.
[c]Efficacy against various pest species within this group varies among nematode species.

Table 1. Use of entomopathogenic nematodes as biological control agents[a] [37].

2.1. Efficacy of entomopathogenic nematodes against tomato leaf miner *Tuta absoluta*

In our laboratory, we investigated the use of native EPN isolates to control various pest insects, and one of these pests was tomato leaf miner. The tomato leafminer, *T. absoluta* (Meyrick) (Lepidoptera: Gelechiidae), is a very devastating pest and was first recorded in 2009 in the Urla District of Izmir Province in Turkey [38]. It has been a serious problem to tomato production in Çanakkale since the first detection in our country [39]. *T. absoluta* can attack all parts and stages of the tomato plant, overwinter in the egg, pupal or adult stage and can cause up to 100% losses in tomato crops [40].

Since its dispersal in the 1970s, chemical control has been the main method to control *T. absoluta*. Producers have tried to decrease its damages by using insecticides twice a week during a cultivation period, sometimes every 4–5 days/season with 8–25 sprays [41]. Although with the many applications of chemicals, effective control is difficult due to the behaviour of these mine-feeding larvae.

Moreover, the use of pesticides in plant production has numerous disadvantages as pesticide residues on human health and on the environment so biological control may be considered as an alternative method to chemical control [42]. In this respect, EPNs can be an alternative to chemicals. The aims of the work were to determine the efficacy of native EPN isolates against *T. absoluta* in tomato field and to reduce the use of pesticides.

2.2. Materials and methods

2.2.1. Entomopathogenic nematodes culture

Four native species of nematodes: *Steinernema affine* (Bovien) (isolate 46) *S. carpocapsae* (Weiser) (isolate 1133), *S. feltiae* (Filipjev) (isolate 879) and *H. bacteriophora* (Poinar) (isolate 1144), were tested against *T. absoluta* larvae. Each isolates was reared in the last instar of wax moth larvae *Galleria mellonella* L., which is the most commonly used insect host for in vivo production of EPNs because of its rich nutrient source available in body and easy to multiply in economical diet source [43, 44].

Nematode-infected *G. mellonella* larvae were placed on white traps [45] at 25°C and IJs that emerged from cadavers were harvested.

2.2.2. Tuta absoluta culture

Larvae, pupae and adults of *T. absoluta* used in the trials were obtained from infested tomato fields in Çanakkale. They reared in wooden rearing cages (50 × 50 × 50 cm) on tomato plants at 25 ± 1°C, 65 ± 5% RH, with a 16:8 L:D photoperiod in climate room.

2.2.3. Field trials

Field trials were carried out in the training and research area of Agriculture Faculty in Çanakkale between 2012 and 2013. In both seasons, nearly 1000 m^2 area was cultivated with tomato and seedlings were controlled periodically and closed by a cage when they reached 20 cm height. Each tomato plant was grown in a single cage (50 × 50 × 50 cm). After 30 days, two males and two females were put into each cage.

EPNs were applied at dusk to utilise the higher air humidity for the nematodes with a conventional airblast sprayer at a rate of 50 IJs/cm^2. Tomato plants remained wet in cages after application for 2 hours and that provides EPNs enough time with perfect condition to find and infect the target pest. The experiment was carried out with two replicates per nematode species and exposure day and repeated twice.

After releasing the adults of *T. absoluta*, EPNs were sprayed on tomato plants at the 7th, 14th and 21st days. Tomato plants were cut from the soil line at the 3rd, 5th 7th, 9th, 11th, 13th and 15th days after EPN applications and analysed to determine the mortality of *T. absoluta*. Dead *T. absoluta* larvae were immediately dissected and checked for nematode infection (**Figure 2**). EPNs most likely entered feeding canals in the leaves of tomatoes. Many larvae of *T. absoluta* died inside these galleries, which indicate that IJs were able to find and infect them.

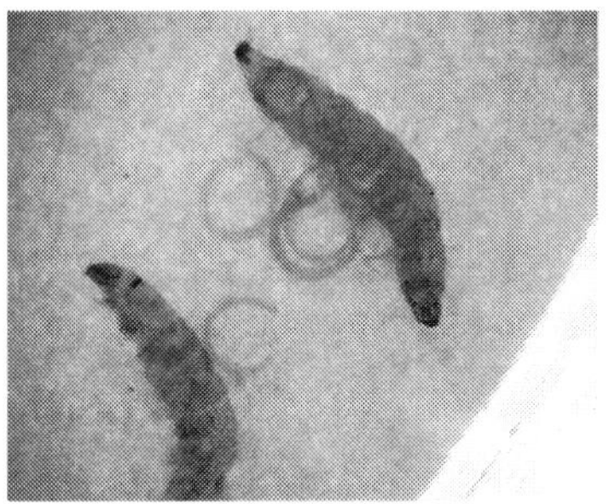

Figure 2. Emerged EPNs from infected *Tuta absoluta* larvae.

2.3. Results

The efficacy of EPNs in field in 2012 changed between 0 and 90.7 ± 1.5%. The least efficient species was *S. affine* and the most efficient species was *S. feltiae* with the mortality of 39.3 ± 1.5% and 90.7 ± 1.5%, respectively. *S. affine* caused 0–39.3 ± 1.5% mortality and found as the least efficient species. *S. carpocapsae* caused 0–43.7 ± 1.5% mortality, while *S. feltiae* caused 0–90.7 ± 1.5% mortality. *H. bacteriophora* caused 0–81 ± 3.5% mortality and was the second efficient species after *S. feltiae* against *T. absoluta* in tomato field in 2012.

The efficacy of EPNs in field in 2013 changed between 0 and 94.3 ± 2.0%. The least efficient species was *S. affine* and the most efficient species was *S. feltiae* with the mortality of 43.7 ± 2.3% and 94.3 ± 2.0%, respectively. *S. affine* caused 0–43.7 ± 2.3% mortality and was the least efficient species. *S. carpocapsae* caused 0–49.3 ± 2.4% mortality, while *S. feltiae* caused 0–94.3 ± 2.0% mortality. *H. bacteriophora* caused 0–83.0 ± 2.1% mortality and was the second efficient species after *S. feltiae* against *T. absoluta* in field in 2013.

2.4. Discussion

The tomato leafminer, *T. absoluta,* is one of the most important lepidopteran moth associated with tomato plants and because of its biology and behaviour, it is difficult to control. Effective chemical control of *T. absoluta* is not possible because it feeds internally within the plant tissues. Resistance to insecticides is another significant problem in chemical control of this pest because of its high reproduction capacity, short generation cycle and intensive use of insecticides [46–50].

Pesticides are so widely used and that destroys populations of natural enemies and consequently decreases biological control of *T. absoluta.* Because of these negative effects of insecticides, other approaches need to be considered seriously for this devastating pest.

Some insects can be controlled by a combination of methods, which are not totally effective when used alone. *T. absoluta* is one of these insects, which requires more than one method to be controlled successfully. For this reason, integrated pest management (IPM) programmes are continuously being progressed in different countries to control infestations of tomato leaf miner. EPNs have been considered as potential biocontrol agents for leafminers in recent years [50]. They can be applied, in combination with other biological and chemical pesticides, fertilisers and soil amendments and in the form of adjuvants or antidesiccants [51, 52].

Various studies about EPNs have been conducted all over the world, but only few research has been carried out on the efficacy of EPNs against *T. absoluta.* This is the first study conducted both in çanakkale and in Turkey based on the efficacy of native EPN isolates to *T. absoluta* in a tomato field.

The efficacy of the three EPNs after foliar application to potted tomato was tested under greenhouse conditions. High larval mortality (78.6–100%) and low pupal mortality (<10%) in laboratory were reported. In the leaf bioassay, high larval parasitisation (77.1–91.7%) was recorded. In the pot experiments, it was found that nematode application decreased insect infestation of tomato by 87–95%. These results showed the suitability of EPNs to control *T. absoluta* [53].

The efficacy of soil treatments of three native EPNs (*S. carpocapsae, S. feltiae* and *H. bacteriophora*) against *T. absoluta* larvae, pupae and adults was determined under laboratory conditions in another study [54]. The effect of three commonly used insecticides against *T. absoluta* was also evaluated in the survival, infectivity and reproduction of these EPNs. When the larvae dropped into the soil to become pupa, soil application of nematodes resulted in a high larval mortality: 100, 52.3 and 96.7% efficacy for *S. carpocapsae, S. feltiae* and *H. bacteriophora,* respectively. No mortality of pupae was recorded, and mortality of adults emerging from soil was 79.1% for *S. carpocapsae* and 0.5% for *S. feltiae.* An insignificant effect of the insecticides tested was reported on nematode survival, infectivity and reproduction. No sublethal effects were observed. These findings proved that larvae of *T. absoluta,* falling from leaves following insecticide application, could be favourable hosts for nematodes, thereby increasing their concentration and persistence in the soil.

The efficacy of *S. feltiae, S. carpocapsae* and *H. bacteriophora* was evaluated against larvae of *T. absoluta* inside leaf mines in tomato leaf discs by means of an automated spray boom. They reported that all EPNs used in the study were effective to all four larval instars of *T. absoluta* but caused higher mortality in the later instars (fourth instar: 77.1–97.4%) than in the first instars (36.8–60.0%). *S. feltiae* and *S. carpocapsae* showed better results than *H. bacteriophora*. *S. carpocapsae* and *H. bacteriophora* performed better at 25°C (55.3 and 97.4% mortality, respectively) than at 18°C (12.5 and 34.2% mortality, respectively), while *S. feltiae* caused 100% mortality at both temperatures. Their results demonstrated that under laboratory conditions, *S. feltiae* and *S. carpocapsae* showed effective performance against the larvae of *T. absoluta* inside tomato leaf mines [55].

Our results agree with other reports showing that larvae of *T. absoluta* were highly susceptible to the EPNs tested and these EPNs can be used as efficient biological control agents against *T. absoluta*. All EPNs used in the study showed efficacy at different rates against *T. absoluta*. They were able to find and infect *T. absoluta* larvae both inside and outside of the tomato leaf. According to these findings, it could be suggested that EPNs have a great potential to use as biocontrol agents for the management of *T. absoluta*.

It should be noted that to understand their life cycles and functions, match the correct species of EPNs with the correct species of insect pests, apply them under optimum environmental conditions, such as soil temperature, soil moisture, angle of sun rays, and apply only with compatible pesticides are the keys to success with EPNs.

3. Conclusions

Biological control is an action that involves the use of natural enemies of insect pests to increase negative effects of insect pest as destroying important crops and plantation, plant growth destruction or development infections caused by pests [56].

Advantages	Disadvantages
Broad host range of pest insect	High cost in production
Able to seek or ambush the host and can kill rapidly the host	Lack of labour, knowledge and skills required in nematology
Mass produced by *in vivo* and *in vitro* (solid and liquid culture medium)	Limited shelf life and refrigerated storage required
Can be used with conventional application equipment	Difficulties in formulation and quality control
Safety for all vertebrates, most non-target invertebrates and the food sources	Environmental limitations; for survival and infectivity adequate moisture and temperatures are needed, sensitivity to UV radiation, lethal effect of several pesticides (nematicides, fumigants and others) lethal or restrictive soil properties (high salinity, high or low pH, etc.)
Little or no registration required	

Table 2. Advantages and disadvantages of entomopathogenic nematodes [58].

EPNs are a group of soil-dwelling organisms that attack soilborne insect pests that live in, on or near the soil surface and can be used effectively to control economically important insect pests. Different nematode species and strains exhibit differences in survival, search behaviour and infectivity, which make them more or less suitable for particular insect pest control programmes [57]. As the other biological control agents, also EPNs have advantages and disadvantages (**Table 2**).

There is a great interest in finding wild populations to obtain new species and strains for possible use in biological control. The use of EPNs is one potential non-chemical approach to control insect pests. EPNs are widely spread geographically and have many hosts. They are currently used as biological control agents in many studies to control several important insect pests worldwide [59–61].

It is highlighted that there is a need for more in-depth basic information on EPNs biology, including ecology, behaviour and genetics, to help understand the underlying reasons for their successes and failures as biological control organisms. Most appropriate nematode species/strain, abiotic factors such as soil type, soil temperature and moisture are important for getting success [1].

Proper match of the nematode to the host entails virulence, host finding and ecological factors are essential before application to the field. Matching the appropriate nematode host-seeking strategy with the pest is essential, because poor host suitability has been the most common mistake occurred in application of EPNs [62]. Also application strategies, such as field dosage, volume, irrigation and appropriate application methods, are very important. Furthermore, plant morphology and phenology must be considered in predicting whether nematodes are viable control candidates [63].

Author details

Ugur Gozel* and Cigdem Gozel

*Address all correspondence to: ugozel@comu.edu.tr

Canakkale Onsekiz Mart University, Agriculture Faculty Plant Protection Department, Çanakkale, Turkey

References

[1] Kaya H.K., Gaugler R. 1993. Entomopathogenic nematodes. Annual Review of Entomology, 38: 181–206.

[2] Laznik Ž., Tóth T., Lakatos T., Vidrih M., Trdan S. 2010. Control of the Colorado potato beetle (*Leptinotarsa decemlineata* [Say]) on potato under field conditions: a comparison

of the efficacy of foliar application of two strains of *Steinernema feltiae* (Filipjev) and spraying with thiametoxam. Journal of Plant Disease Protection, 117: 129–135.

[3] Nickle W.R. 1984. History, development, and importance of insect nematology. In: Nickle W.R. (Ed.) Plant and Insect Nematodes. New York: Marcel Dekker, pp. 627–653.

[4] Poinar G.O. 1990. Taxonomy and biology of Steinernematidae and Heterorhabditidae. In: Gaugler R., Kaya H.K. (Eds.) Entomopathogenic Nematodes in Biological Control, Boca Raton (FL): CRC Press, pp. 23–60.

[5] Nguyen K.B., Smart Jr. G.C. 1994. *Neosteinernema longicurvicauda* n. gen., n. sp. (Rhabditida: Steinernematidae), a parasite of the termite *Reticulitermes flavipes* (Koller). Journal of Nematology, 26: 162–174.

[6] Poinar G.O. Jr. 1976. Description and biology of a new insect parasitic rhabditoid, *Heterorhabditis bacteriophora* n. gen. n. sp. (Rhabditida; Heterorhabditidae n. fam.). Nematologica, 21: 463–470.

[7] Forst S., Dowds B., Boemare N.E., Stackebrandt E. 1997. *Xenorhabdus* spp. and *Photorhabdus* spp.: bugs that kill bugs. Annual Review of Microbiology 51: 47–72.

[8] Abdel-Razek A.S. 2003. Pathogenic effects of *Xenorhabdus nematophilus* and *Photorhabdus luminescens* (Enterobacteriaceae) against pupae of the diamondback moth, *Plutella xylostella* (L.). Anzeiger für Schädlingskunde, 76: 108–111.

[9] Stock S.P. 1995. Natural populations of entomopathogenic nematodes in the Pampean region of Argentina. Nematropica 25: 143–148.

[10] Hominick W.M. 2002. Biogeography. In: Gaugler R. (Ed.) Entomopathogenic Nematology. Wallingford, UK: CABI Publishing, pp. 115–143.

[11] Hominick W.M., Briscoe B.R. 1990. Occurrence of entomopathogenic nematodes (Rhabditida: Steinernematidae and Heterorhabditidae) in British soil. Parasitology, 100: 295–302.

[12] Griffin C.T., Moore J.F., Downes M.J. 1991. Occurrence of insect-parasitic nematodes (Steinernematidae, Heterorhabditidae) in the Republic of Ireland. Nematologica, 37: 92–100.

[13] Stock S.P., Pryor B.M., Kaya H.K. 1999. Distribution of entomopathogenic nematodes (Steinernematidae and Heterorhabditidae) in natural habitats in California, USA. Biodiversity and Conservation, 8: 535–549.

[14] Nguyen K.B., Buss E.A. 2011. *Steinernema phyllophagae* n. sp. (Rhabditida: Steinernematidae), a new entomopathogenic nematode from Florida, USA. Nematology, 13: 425–442.

[15] Malan A.P., Knoetze R., Tiedt L. 2012. *Heterorhabditis noenieputensis* n. sp. (Rhabditida: Heterorhabditidae), a new entomopathogenic nematode from South Africa. Journal of Helminthology 12: 1–13.

[16] Orozco R.A., Hill T., Stock S.P. 2013. Characterization and phylogenetic relationships of *Photorhabdus luminescens* subsp. *sonorensis* (gamma-Proteobacteria: Enterobacteriaceae), the bacterial symbiont of the entomopathogenic nematode *Heterorhabditis sonorensis* (Nematoda: Heterorhabditidae). Current Microbiology, 66: 30–39.

[17] Plichta K.L., Joyce S.A., Clarke D., Waterfield N., Stock S.P. 2009. *Heterorhabditis gerrardi* n. sp (Nematoda: Heterorhabditidae): the hidden host of *Photorhabdus asymbiotica* (Enterobacteriaceae: Gamma-Proteobacteria). Journal of Helminthology, 83: 309–320.

[18] Edgington S., Buddie A.G., Moore D., France A., Merino L., Hunt D.J. 2011. *Heterorhabditis atacamensis* n. sp (Nematoda: Heterorhabditidae), a new entomopathogenic nematode from the Atacama Desert, Chile. Journal of Helminthology, 85: 381–394.

[19] Griffin C.T., O'callaghan K.M., DIX I. 2001. A self-fertile species of *Steinernema* from Indonesia: further evidence of convergent evolution amongst entomopathogenic nematodes? Parasitology 122: 181–186.

[20] Bedding R.A., Molyneux A.S. 1982. Penetration of insect cuticle by infective juveniles of *Heterorhabditis* spp. (Heterorhabditidae: Nematoda). Nematologica, 21: 109–110.

[21] Grewal P.S., Lewis E.E., Gaugler R., Campbell J.F. 1994. Host finding behavior as a predictor of foraging strategy in entomopathogenic nematodes. Parasitology, 108: 207–215.

[22] Kung S.P., Gaugler R., Kaya H.K. 1991. Effects of soil temperature, moisture, and relative humidity on entomopathogenic nematode persistence. Journal of Invertebrate Pathology, 57: 242–249.

[23] Campbell J.F., Lewis E.E., Stock S.P., Nadler S., Kaya H.K. 2003. Evolution of host search strategies in entomopathogenic nematodes. Journal of Nematology, 35: 142–145.

[24] Grewal P.S., Ehlers R-U., Shapiro-Ilan D.I. 2005. Nematodes as Biological Control Agents. Wallingford: CABI Publishing.

[25] Kaya H.K. 1990. Soil ecology. In: Gaugler R., Kaya H.K. (Eds.) Entomopathogenic Nematodes in Biological Control. Boca Raton (FL): CRC Press, pp. 93–115.

[26] Ellsbury M.M., Jackson J.J., Woodson W.D., Beck D.L., Stange K.A. 1996. Efficacy, application distribution, and concentration by stemflow of *Steinernema carpocapsae* (Rhabditida: Steinernematidae) suspensions applied with a lateral-move irrigation system for corn rootworm (Coleoptera: Chrysomelidae) control in maize. Journal of Economic Entomology, 85: 2425–2432.

[27] Koppenhofer A.M., Fuzy E.M. 2006. Nematodes for white grub control: Effects of soil type and soil moisture on infectivity and persistence. USGA Turfgrass Environmental Research Online, 5: 1–10.

[28] Lacey L.A., Georgis R. 2012. Entomopathogenic nematodes for control of insect pests above and below ground with comments on commercial production. Journal of Nematology, 44(2): 218–225.

[29] Poinar G.O. Jr., Thomas G.M., Presser S.B., Hardy J.L. 1982. Inoculation of entomogenous nematodes, *Neoaplectana* and *Heterorhabditis* and their associated bacteria, *Xenorhabdus* spp., into chicks and mice. Environmental Entomology, 11: 137–138.

[30] Boemare N.E., Laumond C., Mauleon H. 1996. The entomopathogenic nematode-bacterium complex: biology, life cycle and vertebrate safety. Biocontrol Science Technology, 6: 333–346.

[31] Poinar Jr G.O., Thomas G.M. 1988. Infection of frog tadpoles (Amphibia) by insect parasitic nematodes (Rhabditida). Experientia, 44: 528–531.

[32] Kermarrec A., Mauleon H., Sirjusingh C., Bauld L. 1991. Experimental studies on the sensitivity of tropical vertebrates (toads, frogs and lizards) to different species of entomoparasitic nematodes of the genera *Heterorhabditis* and *Steinernema*. Rencontres Caraibes Lutte Biologique. In Proceedings of the Caribbean Meetings on Biological Control, Guadeloupe, pp. 193–204.

[33] Georgis R., Kaya H.K., Gaugler R. 1991. Effect of Steinernematid and Heterorhabditid nematodes (Rhabditida: Steinernematidae and Heterorhabditidae) on non-target arthropods. Environmental Entomology, 20: 815–822.

[34] Bathon H. 1996. Impact of entomopathogenic nematodes on non-target hosts. Biocontrol Science and Technology, 6: 421–434.

[35] Griffin C.T., Downes M.J., Block W. 1990. Test of Antarctic soils for insect parasitic nematodes. Antarctic Science, 2: 221–222.

[36] Akhurst R., Smith K. 2002. Regulation and safety. In: Gaugler R. (Ed.) Entomopathogenic Nematology. Wallingford, UK: CABI Publishing, pp. 311–332, 400 pp.

[37] Shapiro-Ilan, D.I., R. Gaugler. 2010. Nematodes: Rhabditida: Steinernematidae & Heterorhabditidae. In: Shelton A. (Ed.) Biological Control: A Guide to Natural Enemies in North America. Cornell University. http://www. biocontrol.entomology.cornell.edu/pathogens/ nematodes.html.

[38] Kılıç T. 2010. First record of *Tuta absoluta* in Turkey. Phytoparasitica, 38: 243–244.

[39] Kasap İ., Gözel U., Özpınar A. 2011. A new pest in tomatoes; the tomato borer, Tuta absoluta (Meyrick) (Lepidoptera: Gelechiidae). Proceedings of Çanakkale Agriculture Symposium (Yesterday, Today, Future), Çanakkale Onsekiz Mart University, Agriculture Faculty, Çanakkale, 284–287.

[40] EPPO, 2005. Data sheets on quarantine pests: *Tuta absoluta*. OEPP/EPPO Bulletin, 35: 434–435.

[41] Temerak S.A. 2011. The status of *Tuta absoluta* in Egypt, 16–18. EPPO/IOPC/FAO/NEPP Joint, International Symposium on Management of *Tuta absoluta* (tomato borer), Conference, Agadir, Morocco, November, 18 pp.

[42] Gill H.K., Garg H. 2014. Pesticide: environmental impacts and management strategies. In: Solenski S., Larramenday M.L. (Eds.) Pesticides-Toxic Effects. Rijeka, Croatia: Intech, pp. 187–230.

[43] Bedding R.A., Akhurst R.J., 1975. A simple technique for the detection of insect parasitic Rhabditid nematodes in soil. Nematologica, 21: 109–110.

[44] Kaya H.K., Stock S.P. 1997. Techniques in insect nematology, In: Lacey L.A. (Ed.) Manual of Techniques in Insect Pathology. Biological Techniques Series, San Diego, California: Academic Press, pp. 281–324. 409 pp.

[45] White G.F. 1927. A method for obtaining infective nematode larvae from cultures. Science, 66: 302–303.

[46] Salazar E.R., Araya J.E. 1997. Detection of resistance to insecticides in the tomato moth. Simiente, (1–2): 8–22.

[47] Salazar E.R., Araya J.E. 2001. Tomato moth, Tuta absoluta (Meyrick) response to insecticides in Arica, Chile. 61(4): 429–435.

[48] Siqueira H.A.A., Guedes R.N.C., Picanço M.C. 2000. Insecticide resistance in populations of *Tuta absoluta* (Lepidoptera: Gelechiidae). Agricultural and Forest Entomology, 2: 147–153.

[49] Siqueira, H.A.A., Guedes R.N.C., Fragoso D.B., Magalhaes L.C. 2001. Abamectin resistance and synergism in Brazilian populations of *Tuta absoluta* (Meyrick) (Lepidoptera: Gelechiidae). International Journal of Pest Management, 47(4): 247–251.

[50] Olthof Th. H.A., Broadbent A.B. 1990. Control of a chrysanthemum leafminer, *Liriomyza trifolii* with the entomophilic nematode, *Heterorhabditis heliothidis*, 379. Abstracts of Papers, Posters and Films Presented at the Second International Nematology Congress, 11–17 August, Veldhoven the Netherlands, Nematologica, 36(1990): 327–403.

[51] Glazer I., Navon A. 1990. Activity and persistence of entomogenous nematodes used against *Heliothis armigera* (Lepidoptera: Noctuidae). Journal of Economic Entomology, 83: 1795–1800.

[52] Baur M.E., Kaya H.K., Gaugler R., Tabashnik B. 1997. Effects of adjuvants on entomopathogenic nematode persistence and efficacy against *Plutella xylostella*. Biocontrol, Science and Technology, 7: 513–525.

[53] Batalla-Carrera L., Morton A., Garcia-del-pino F. 2010. Efficacy of entomopathogenic nematodes against the tomato leafminer *Tuta absoluta* in laboratory and greenhouse conditions. Biocontrol, 55: 523–530.

[54] Garcia del Pino F., Alabern X., Morton A. 2013. Efficacy of soil treatments of entomopathogenic nematodes against the larvae, pupae and adults of *Tuta absoluta* and their interaction with the insecticides used against this insect. BioControl, 58(6): 723–731.

[55] Van Damme V.M., Beck B.K., Berckmoes E., Moerkens R., Wittemans L., Vis R.D., Nuyttens D., Casteels H.F., Maes M., Tirry L., Clercq P.D. 2015. Efficacy of entomopathogenic nematodes against larvae of *Tuta absoluta* in the laboratory. Pest Management Science, doi:10.1002/ps.4195, 2015.

[56] Flint M.L., Dreistadt S.H. 1998. Natural Enemies' Handbook: The Illustrated Guide to Biological Pest Control. Berkeley, CA: University of California Press, pp. 2–35.

[57] Garcia del Pino F., Palomo A., 1996. Natural occurrence of entomopathogenic nematodes (Rhabditida: Steinernematidae and Heterorhabditidae) in Spanish soil. Journal of Invertebrate Pathology, 68: 84–90.

[58] Shapiro-Ilan D.I., Bruck D.J., Lacey L.A. 2012. Principles of epizootiology and microbial control. In Vega F.E., Kaya H.K., (Eds.) Insect Pathology, Second Ed. San Diego: Academic Press, pp. 29–72.

[59] Shields E.J., Testa A., Miller J.M., Flanders K.L. 1999. Field efficacy and persistence of the entomopathogenic nematodes *Heterorhabditis bacteriophora* 'Oswego' and *H. bacteriophora* 'NC' on Alfalfa snout beetle larvae (Coleoptera: Curculionidae). Environmental Entomology, 28: 128–136.

[60] Gozel U., Gunes C. 2013. Effect of entomopathogenic nematode species on the corn stalk borer (*Sesamia cretica* Led. Lepidoptera: Noctuidae) at different temperatures. Turkish Journal of Entomology, 37(1): 65–72.

[61] Gozel C., Kasap İ. 2015. Efficacy of entomopathogenic nematodes against the Tomato leafminer, *Tuta absoluta* (Meyrick) (Lepidoptera: Gelechiidae) in tomato field. Turkish Journal of Entomology, 39(3): 229–237.

[62] Gaugler R. 1999. Matching nematode and insect to achieve optimal field performance. In Polavarapu S. (Ed.) Workshop Proceedings: Optimal use of Insecticidal Nematodes in Pest Management, Rutgers University, pp. 9–14.

[63] Georgis R., Koppenhöfer A.M., Lacey L.A., Bélair G., Duncan L.W., Grewal P.S., Samish M., Tan L., Torr P., van Tol R.W.H.M. 2006. Successes and failures in the use of parasitic nematodes for pest control. Biological Control, 38: 103–123.

Chapter 4

Novelties in Pest Control by Entomopathogenic and Mollusc-Parasitic Nematodes

Vladimír Půža, Zdeněk Mráček and Jiří Nermuť

Additional information is available at the end of the chapter

http://dx.doi.org/10.5772/64578

Abstract

Entomopathogenic and molluscoparasitic nematodes are important parasites of many insects and molluscs, respectively. Due to their infectivity, the possibility of mass production by industrial techniques and the relative safety to nontarget organisms and environment, these organisms represent an attractive agent for biological control of many pests. This chapter summarises the current knowledge of the diversity of these organisms. In this chapter, we review the recent advances in production, storage, application techniques genetic improvement and safety of these organisms.

Keywords: nematodes, entomopathogenic, molluscoparasitic, *Steinernema*, *Heterorhabditis*, *Phasmarhabditis*, diversity, occurrence, rearing, application, safety

1. Introduction

Entomopathogenic nematodes (EPNs) in the families of *Steinernematidae* and *Heterorhabditidae* are important parasites of many insect species. Due to their ability to infect various insects, the possibility of mass production by industrial techniques and the relative safety to nontarget organisms and environment, EPNs represent an attractive agent for biological control of many insect pests.

Over the past decade, a large number of new EPN species have been described from throughout the world. New lineages present a unique combination of characteristics and thus have a great potential for biological control of particular insect pests.

Mollusc-parasitic nematodes (MPNs) represent a taxonomically more diverse group, consisting of members of seven families (Agfidae, Alaninematidae, Alloionematidae, Angiostomati-

dae, Cosmocercidae, Diplogasteridae, Mermithidae and Rhabditidae). However, to date, only *Phasmarhabditis hermaphrodita* (Rhabditidae) has been commercialised. It is likely that other mollusc-parasitic nematodes have a potential to provide new bio-agents for slug and snail control. MPN biology is mostly unknown, but recently published descriptions of several new species provided at least some notes about MPN biology.

This chapter provides thorough information about the diversity and biology of EPNs and MPNs. We also focus on the recent advances in production, storage and application techniques.

2. Overview of EPN and MPN biology and diversity

2.1. Diversity of entomopathogenic nematodes

EPNs are common in all types of soils and more frequently inhabit agricultural and secondary forest ecosystems, which represent suitable conditions for insect host populations. These organisms have a worldwide distribution [1]. Over the past few decades, numerous surveys were performed mainly in Europe [2] and North America [3]. However, recently a huge effort for the study of EPNs field occurrence was recorded from other continents of all zoogeographical regions. Results of these surveys increased rapidly a number of new described species, especially from South Africa, Ethiopian region [4], Southeast Asia, Indo-Malaysian region [5] and tropical areas in Neotropical region [6]. Over the past decade, regularly used DNA analysis facilitated discrimination of the morphologically almost identical sibling species.

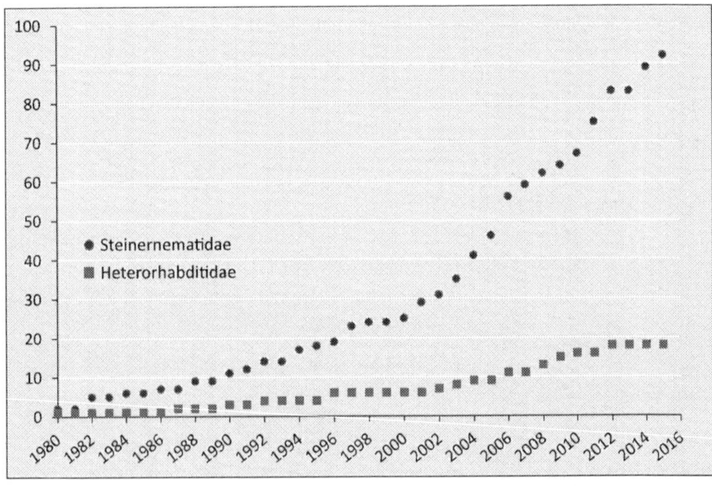

Figure 1. The increasing number of recognised steinernematid and heterorhabditid species based on published data.

This led to the tremendous increase in the known EPN diversity. From the year 2000, the number of the described steinernematids and heterorhabditids more than tripled from 25 to 92 and from 6 to 18, respectively (**Figure 1**). Understanding of EPNs diversity should be considered as a basic requirement for a successful field control of noxious insects.

It is generally accepted that steinernematids are more common in cooler, temperate zone whereas heterorhabditids prefer warmer, tropical and subtropical conditions (torrid zone) [7]. Geographically, the torrid zone lies between the Tropic of Capricorn and the Tropic of Cancer parched with heat. In this zone, many new species have been recently detected from Vietnam and southern China. Temperate zones contain the areas or regions between the tropic of Capricorn and the Antarctic circle or between the tropic of Cancer and the arctic circle, having a moderate climate. According to the number of described species, this zone seems to be the richest for the EPNs occurrence. Frigid zones represent the areas or regions between the Antarctic circle and the south pole or between the arctic circle and north pole, intensively cold, have probably a low EPNs occurrence represented only by several findings. *Steinernema kraussei* and recently *Steinernema affine* are the only species with a link to the frigid zone [8, 9].

Continent	*Steinernema*/ *Heterorhabditis*	From 2010	Species
Africa	16/6	11	*S. citrae, S. cameroonense, S. ethiopiense, S. nyetense, S. sacchari, S. innovationi, S. tophus, S. jeffreyense, S. fabii, S. pwaniensis, H. noenieputensis*
Asia	52/5	14	*S. minutum, S. nepalense, S. surkhetense, S. everestense, S. lamjungense, S. pui, S. changbaiense, S. tielingense, S. xinbinense, S. dharanaii, S. bifurcatum, S. huense, S. balochiense, H. beicherriana*
Australia	3/4	0	–
Europe	16/4	3	*S. schliemanni, S. vulcanicum, S. poinari*
North and Central America	15/9	1	*S. phyllophagae*
South America	15/4	5	*S. brazilense, S, unicronum, S. papillatum, S. goweni, H. atacamensis*

Table 1. Number of steinernematid and heterorhabditid species by continent and number and identity of the EPN species described from each continent since the year 2010 based on published data.

The highest species diversity of the genus *Steinernema* is found in the Asian continent, with 52 recorded species, whereas the area of North and Central America has the highest number of heterorhabditids with 9 recorded species (**Table 1**). The Asian and African continents are also the areas with the fastest growing numbers of the described EPNs with 11 and 14 described EPNs since the year 2010. Europe has the longest tradition of EPN research and is the most extensively and intensively sampled continent. Despite this fact, three new steinernematids have been recovered in the past 5 years. This fact suggests that we are likely to see much more new EPNs to be described from other continents in the future.

2.1.1. Geographic distribution of EPN species

Several EPNs are known to have a cosmopolitan occurrence, such as *S. kraussei, Steinernema glaseri, Steinernema feltiae, Steinernema carpocapsae, Heterorhabditis bacteriophora, Heterorhabditis indica, Heterorhabditis megidis* and *Heterorhabditis zealandica* (**Table 2**). Among them, we can distinguish those that prefer temperate or torrid zone, or occur in both these zones. Of these, *S. kraussei* is a Holoarctic species and its recovery from Neotropic in Colombia is a unique observation or doubtful result [10]. Similarly, *S. glaseri* and *S. carpocapsae* inhabit preferably Holoarctic temperate zone with links to torrid zone in Indo-Malaysian (India/Tamil Nadu) and Neoarctic (SE USA) regions [5, 11]. *S. feltiae* seems to be the best adapted species inhabiting all continents, warm and cool areas and wide spectrum of habitats. Surely, this is the most common steinernematid in Holoarctic and Neotropic and Australian regions, recently found also in Indo-Malaysian [12] and Afrotropical [13] regions. *H. bacteriophora*, originally described from Australian region, is the most widespread heterorhabditid. The nematode occurs in all zoogeographical regions including both torrid and temperate zones. *H. indica* is the nematode widespread over the torrid zone in tropical and subtropical areas of all zoogeographical regions, whereas *H. megidis* has been discovered only in temperate zone of Holoarctic. An interesting distribution is reported for *H. zealandica*, originally described from New Zealand, later found in the northeastern Europe. This species was recently reported also from north-eastern China [14], Florida [15] and, surprisingly, also from South Africa [4].

Species	Distribution
S. carpocapsae	Worldwide, all continents
S. feltiae	Worldwide, all continents
H. bacteriophora	Worldwide, all continents
H. indica	Worldwide, all continents
H. zealandica	Worldwide, Australia, Africa, North America, Europe
S. glaseri	Holarctic—USA, Argentina, Azores, China, Korea and Spain
S. affine	Holarctic—Europe, Russia, Canada
S. kraussei	Holarctic—Europe, Russia, Canada
S. arenarium	Palearctic—Europe, Russia
S. intermedium	Palearctic—USA, Europe
S. poinari	Palearctic—Europe, Russia
H. megidis	Palearctic—North America, Asia, Europe
S. abassi	Northern Africa, India
S. weiseri	Central and Northern Europe, Turkey
S. yirgalemense	Central and Southern Africa
S. silvaticum	Central and Northern Europe, United Kingdom

Table 2. Entomopathogenic nematode species with a large geographic range and their distribution.

Bacterium	Nematode
X. bovienii	*S. affine, S. jollieti, S. feltiae, S. cholashanense, S. ichnusae, S. intermedium, S. kraussei, S. sichuanense, S. weiseri, S. xueshanense*
X. budapestensis	*S. bicornutum, S. ceratophorum*
X. beddingi	Unknown
X. cabanillasii	*S. riobrave*
X. doucetiae	*S. diaprepesi*
X. ehlersii	*S. longicaudum*
X. griffiniae	*S. hermaphroditum*
X. hominickii	*S. karii, S. monticolum*
X. indica	*S. yirgalemense, S. abassi*
X. innexi	*S. scapterisci*
X. ishibashi	*S. aciari*
X. japonica	*S. kushidai*
X. khoisanae	*S. khoisanae, S. pwaniensis*
X. koppenhoeferi	*S. scarabei*
X. kozodoii	*S. apuliae, S. arenarium*
X. magdalenensis	*S. australe*
X. mauleonii	Unknown
x. miraniensis	Unknown
X. nematophila	*S. carpocapsae*
X. poinarii	*S. cubanum, S. glaseri*
X. romanii	*S. puertoricense*
X. stockiae	*S. huense, S. minutum, S. siamkayai*
X. szentirmaii	*S. rarum*
X. vietnamensis	*S. sangi*
P. asymbiotica	*H. gerrardi, H. indica*
P. heterorhabditis	*H. zealandica*
P. luminescens	*H. bacteriophora, H. georgiana, H. noenieputensis, H. sonorensis, H. indica*
P. temperata	*H. bacteriophora, H. downesi, H. georgiana, H. megidis*

Table 3. Taxonomic correspondence of symbiotic bacteria of the genera *Xenorhabdus* and *Photorhabdus* to host entomopathogenic nematodes.

In contrast to ubiquitous species, the known geographic distribution of a majority of EPN species is much narrower, and some species are known just from a single country and even from a specific locality. This applies, for instance to *Steinernema vulcanicum*, to date found only in Italian island of Sicily [16]. However, at least in some species, their known geographic range will probably expand with more data available in the future.

It is likely, that at least in some ubiquitous EPNs, their geographical distribution was recently enhanced by a human activity. This seems to be the case of *S. affine* that was known to occur throughout the Europe, and it was believed to have a Palearctic distribution. However, in 2005, it was recovered in British Columbia [17], North America. Such geographic distribution could be either due to its historically forming a disjunctive range of the species in a Holoarctic distribution or, and more likely, *S. affine* has been introduced into Greater Vancouver by immigrants and/or by imported commercial produce, such as potatoes (*Solanum tuberosum*), flower bulbs and other agriculture plants transported from Europe. Following its arrival, it then spread over the Greater Vancouver coastal area. In other species, it is often impossible to imply, whether they are indigenous to a given locality. It can be however assumed, that indigenous species should be considered only those isolated in natural, climax, ecosystems, for example, *Steinernema brazilense* [18].

Unfortunately, the data on EPN diversity is partly influenced by the wrong identification of certain species. For instance, some EPNs originally described from South and North America, such as *Steinernema ritteri, Steinernema rarum, Steinernema scapterisci* and *Steinernema riobrave,* were later reported from Northeast China [14], which is at least doubtful. There are more species with reportedly very disjunctive distribution, overlapping different zoogeographical regions such as *Steinernema bicornutum* described from Serbia but later reported from Jamaica [19].

2.1.2. Habitat preference

EPNs inhabit most terrestrial habitats, but their occurrence has been evaluated mainly in relation to soil type and habitat [20]. Interestingly, heterorhabditids were equally abundant in turf and weedy habitats, but never found in closed-canopy forest [21]. In Germany, the rate of prevalence of steinernematids was highest in woodland (50.3%) where *S. affine, S. feltiae, Steinernema intermedium* and *Steinernema silvaticum* (=*Steinernema* sp. B) were the predominant species [22]. This fact can be explained by the higher insect host occurrence in woodland habitats in comparison with the usually poor field ecosystems. Many field studies solved an impact of various abiotic factors for EPNs recovery, survival etc. The EPN occurrence in Spain was evaluated through abundance, recovery frequency, larval mortality percentage and EPN population density. EPNs occurrence was also related to the selected soil physical and chemical variables as well as to some soil pollutants such as heavy metals and organochlorine pesticide residues. These factors help to understand how EPNs survive and disperse [23], but as usually no data were published about natural insect hosts. Recently, ten species of *Steinernematidae* including three undescribed and three species of Heterorhabditidae confirmed a rich EPNs fauna in northern China. Their occurrence was strongly associated to the prevailing climatic conditions, altitude, vegetation and soil types [24].

In general, the essential condition for the EPN occurrence and survival associates with biotic factors. Different species of EPNs occur in numerous habitats/ecosystems depending preferably on their insect host. It was demonstrated that at least some steinernematids show a distinct habitat preference that may reflect the distribution of suitable hosts, which are adapted for the habitat [7]. Even though, these nematodes are ubiquitous, their recovery from the field is influenced by a number of biotic factors, including nematode antagonists and host range that is dependent on the suitability for penetration of different insect hosts by nematodes, possibility of finding a suitable host in the habitats (e.g., leaf-feeding insects cannot be readily attacked in the natural habitat), and by the natural population density [25]. Till present, the impact of insect hosts has been, unfortunately, mostly overlooked. Insect aggregations and outbreaks of insect pests are a great opportunity to study EPN diversity and habitat preferences. Mráček and Bečvář [26] emphasised an essential impact of host aggregations on the incidence of EPNs. In their experiments, the high percentage, about 70%, of sampling sites with insect aggregations were nematode positive. Similarly, final mortality of the fly larvae and pupae from the bibionid (*Bibio marci*) nest aggregation caused by *S. intermedium* achieved 90% [27]. Even though, occurrence of suitable insect hosts in habitats seems to be elementary for the incidence of EPNs, at least some species are behaviourally adapted for different types of soil and habitat. In general, heterorhabditids prevail in light, sandy soils whereas soil type is less important for steinernematids and *S. kraussei, S. intermedium* and *S. silvaticum* are abundant species in forest habitats.

Competition between EPN species can also have an impact on their distribution. It was shown that even though *Heterorhabditis* and *Steinernema* can co-infect the host, they cannot coexist and one genus will prevail [28]. Two steinernematid species, on the other hand, can co-infect and reproduce within one host cadaver [29]; however, one or both species are often negatively affected by competition [30]. In British Columbia, Canada field sampling identified *S. affine* occurring together with *S. kraussei* at two sites [17]. In the field, the *Galleria* baiting and consequent laboratory experiments, *S. affine* appeared to be a more successful parasite than *S. kraussei*. When *Galleria mellonella* larvae were co-infected by *S. carpocapsae* and *S. glaseri*, the proportion of established females was reduced in cadavers and the progeny of *S. glaseri* was less affected by the mixed infection than that of *S. carpocapsae* [29]. Similarly, the interactions of two sympatric entomopathogenic nematodes, *S. affine* and *S. kraussei,* were studied in a series of laboratory experiments [30]. In the co-infections, *S. kraussei* was strongly negatively affected while *S. affine* was able to multiply in a higher number of hosts in comparison to single infection and it was also able to invade and multiply in hosts already infected and even killed by *S. kraussei* and it produced a normal amount of progeny. The field study in the original locality [31], however, found no spatial relationship between the two species, and no evidence suggesting any host differentiation between the two species was found. Authors assumed that both species share an ecological niche, and thus the avoidance of competition with the latter species seems to be a crucial factor for *S. kraussei*. Patchy distribution and implicit differences in horizontal distribution probably markedly contribute to the coexistence of both species.

2.1.3. Methods used for the study of EPN diversity

The outcome of the studies of EPN occurrence can be influenced by the method of isolation. A total of 40 soil samples from various habitats in Germany and the Czech Republic were studied for the presence of entomopathogenic nematodes using the *Galleria* baiting and a sieving-decanting method for direct extraction of infective-stage juveniles [32]. All these species were recovered with both methods, but the baiting technique was generally less effective, and mixtures of several species in one soil sample were frequently undetected. The direct extraction method provided quantitative estimates of infective stage juvenile density, but no information on their infectivity or on morphological characters of adults and nematode cultures could be established. However, *Galleria* baiting could be negatively influenced by EPNs competition when one species' infective activity can be suppressed by another one [17, 31]. Besides these classical baiting methods, the quantitative real-time PCR (qPCR) techniques have been recently used to provide accurate and reliable methods to identify and quantify cryptic organisms in soil ecology [33]. By this method, six species of EPNs were recovered in Florida citrus (*Citrus* spp.) orchards (*S. glaseri, Steinernema diaprepesi, S. riobrave, H. indica, H. zealandica, Heterorhabditis floridensis* and an undescribed species in the *S. glaseri* group). The qPCR assay was more efficient than the *Galleria* baiting method for detecting the EPN species composition in species mixtures and represents a new challenge for the EPNs biodiversity studies. The classical *Galleria* baiting method uses larvae of the greater wax moth (*G. mellonella*) that are placed to the soil sample and invaded by EPN infective larvae. However, this method can miss inactive or competitively weak EPNs. In the qPCR method, the total DNA is extracted from the soil sample or the infected *Galleria* larva, and EPN species are detected and quantified by qPCR with species-specific probes.

2.1.4. Symbiotic bacteria

Entomopathogenic nematodes (EPNs) of the families *Steinernematidae* and *Heterorhabditidae* are mutualistically associated with specific symbiotic bacteria of the genus *Xenorhabdus* and *Photorhabdus*, respectively [34]. The relationship is obligate in natural environment [35]. Besides providing the food source to the nematodes, bacteria also protect cadaver against other microorganisms by production of bacteriocins, antibiotics and antimicrobials [36, 37] and against insect scavengers [38].

Single species of *Steinernema* may be associated with only one species of *Xenorhabdus*. The same applies to *Heterorhabditis* with the exception of *H. bacteriophora* that is associated either with *Photorhabdus luminescens* or with *Photorhabdus temperata*. On the other hand, species of *Photorhabdus* and certain species of *Xenorhabdus* are hosted by several species of *Heterorhabditis* and *Steinernema* (**Table 3**).

2.2. Molluscs-parasitic nematodes (MPNs)

Nematode parasites of molluscs (mollusc-parasitic nematodes) can be found in several families (e.g., Alloionematidae, Cosmocercidae, Mermithidae and Rhabditidae). Of these, only one species, *P. hermaphrodita* (Rhabditidae), has been commercialised. However, several other mollusc-parasitic nematodes might have a high potential to provide new bio-agents for harmful molluscs control. In the following section, we give an overview of the biology and diversity of MPNs.

http://dx.doi.org/10.5772/64578

Similarly to entomopathogenic nematodes, most of MPNs spend a part of their life cycle in the soil environment. Nematodes that infect host in the soil need some mechanism how to find an appropriate host. Soil dwelling invertebrates movement is usually slow, but still too fast for the nematodes and thus, during their evolution, parasitic nematodes developed useful adaptations. As known from EPNs, also *P. hermaphrodita* [39, 40] but very probably also many other nematodes, readily react to host-associated cues. This can be CO_2 or other volatile compounds produced by the living host, its faeces, mucus, etc. Parasitic nematodes react very strongly to all of them, not only to alive host. Of course it is not surprising statement, we know that *P. hermaphrodita* or other MPNs are able to complete their life cycles on slug faeces and other organic matter [41–43]. This type of behaviour provides the nematode also other advantage. Molluscs show a homing behaviour. They use the same shelters every day or night, and usually they cover this place by a big amount of mucus and faeces very soon. Therefore, the nematodes that readily react to these cues gain an advantage and increase their chance to meet the suitable host. Interesting finding is that *P. hermaphrodita* can strongly react not only to water soluble cues as most of other nematodes do but also to volatile cues [40], which can be related with its habitat, soil surface and organic matter, which is inhabited by its hosts.

Unlike EPNs, some MPNs are able to complete their life cycles in different organic matter in the soil. Naturally, the quality of the growing substrate affects nematode development, however, unlike in EPNs, the quality of the growing substrate is mostly expressed in the yield of dauer juveniles and not in the quality of progeny [42]. On the other hand, in EPNs, the substrate quality influences both yield and quality of IJs [44, 45]. The reason for this difference could be that while EPNs are the true parasites, MPNs retain both parasitic and free-living life cycles, and the ability to produce full quality dauer juveniles in a wide range of conditions is an essential advantage that helps them to survive in various changing environments.

2.2.1. *Alloionematidae*

Family Alloionematidae consists of three genera: *Alloionema* (with only one species: *A. appendiculatum*), *Neoalloionema* and *Rhabditophanes*. *Alloionema appendiculatum* is a common larval parasite of many terrestrial molluscs that was described from the body of slug *Arion ater* [46]. This nematode retains both parasitic and free-living life cycle [41]. Its dauer juveniles (third-stage larvae) invade foot muscle of snails and slugs and after moulting into the fourth-stage larvae, they stay encysted inside the host muscle. These fourth-stage larvae are able to leave slugs to mature and reproduce in the soil (parasitic generation). The progeny of the parasitic generation makes a free-living saprophytic generation that can live in a suitable organic material for very long time, at least 4 years in laboratory conditions [43]. Development of the saprophytic form is fast and the whole life cycle is completed within 72 or 96 h. When the source of food is depleted, new DJs are produced and spread in the soil to infect new hosts. All stages of both saprophytic and parasitic generations are bacteriophagous and freely associated with many bacteria, for example, *Acinetobacters* sp., *Pseudomonas* sp. and *Neisseria* sp. [43]. *A. appendiculatum* parasites in molluscs belonging to the families Agriolimacidae, Arionidae, Helicidae, Hygromiidae and Succineidae and its prevalence ranges from less than 0.01% in *Cantareus aspersus* up to 100% in some arionid slugs [43, 47].

2.2.2. *Cosmocercidae*

The nematodes in the family Cosmocercidae are usually parasites of reptiles and amphibians, but two genera are known as mollusc-parasites, namely *Nemhelix* and *Cosmocercoides*. *Cosmocercoides dukae* parasites in pallial cavity of many North American slugs and snails. *Nemhelix bakeri* and some other species of the genus parasite in reproductive organs of European helicid snails. Under natural conditions, *N. bakeri* is frequently associated with *Helix aspersa*. This nematode lives and reproduces in genital tract of its host. Infection of the new host by *N. bakeri* occurs only during mating, when the parasite is exchanged along with the spermatophores [48]. It means that juvenile molluscs are always free of infection. *N. bakeri* reduces the fecundity of their hosts [49] but their potential for mollusc biocontrol is still questionable.

2.2.3. *Mermithidae*

Mermithids are frequent parasites of many invertebrates in aquatic and terrestrial habitats, for example, *Romanomermis culicivorax* parasiting in mosquito larvae or *Mermis nigrescens* that is quite frequent parasite of grasshoppers and molluscs [50]. Mermithids are commonly found in mollusc hosts, but it seems that they use molluscs only as facultative hosts [47]. *M. nigrescens* has parasitic larvae and free-living adults that lay eggs on plants, especially on leaves, usually early in the morning or in the night, when there is high humidity. Eggs that are very resistant to dry conditions and UV radiation are able to persist on plants for the whole season. When the eggs are eaten by the suitable host, invasive larvae hatch and penetrate into the haemocoel through the gut wall and develop for several weeks. 'Grown up' larva leaves the host by penetrating its body wall. Host is usually infected with some pathogens through the opening and dies shortly afterward. Emerged larvae develop into post parasites in the soil and adults mate later. The whole development in the soil can take several months, and therefore the whole life cycle can take more than 1 year.

2.2.4. *Rhabditidae*

Rhabditidae is a large family consisting of many bacteriophagous free-living, phoretic and parasitic nematodes that are often associated with insects or terrestrial molluscs (e.g., *Rhabditis*, *Caenorhabditis* or *Phasmarhabditis*) and other invertebrates. The slug parasitic nematode *P. hermaphrodita* (Schneider) Andrassy is almost cosmopolitan species capable of infecting many slug and snail species, such as Arionidae, Agriolimacidae or Limacidae. The dauer juveniles (DJs) infect slugs in the area beneath the mantle surrounding the shell. They usually cause a disease with characteristic symptoms, particularly a swelling of the mantle. The infection often leads to the death of the slug, within 1–3 weeks. New DJs, which are released from the host cadaver, spread into the soil and look for new hosts [51]. Apart from the parasitic cycle, *P. hermaphrodita* also has a necromenic life cycle [41] and has been shown to reproduce on dead earthworms [52], leaf litter [53] and slugs or slug faeces [40]. *P. hermaphrodita* does not live in a strict association with only one species of bacteria as EPNs do, but is associated with many bacterial species [54, 55] that are common in its habitat. Bacterial species are responsible for the pathogenicity of nematode-bacteria complex towards their hosts [56].

3. Mass production

3.1. Entomopathogenic nematodes

An excellent review of the current situation regarding mass production of EPNs was published by Shapiro-Ilan et al. [57]. Therefore, in this chapter, we give only a short overview of the used methods.

The most simple method for EPN production is *in vivo* method, using living insects, mostly the greater wax moth (*Galleria mellonella*) larvae, or mealworms (*Tenebrio molitor*) that are both very susceptible to EPN infection, and their bodies contain enough nutrients for EPN reproduction. This method is simple and cheap, but is labour and cost-effective only at a small scale, and it is therefore appropriate for laboratory use or small-scale applications [58].

For large-scale production, solid or liquid fermentation *in vitro* technologies must be used. At first, EPNs were cultured axenically both in solid [59] and liquid media [60]. Nowadays, the nematodes are always cultured monoxenically to ensure quality consistency and predictability [61]. A symbiont is extracted from the nematodes, and subsequently sterile nematode eggs are applied to the medium pre-inoculated with bacterial symbiont.

EPN production in solid culture is usually performed in a three-dimensional rearing system with the liquid medium mixed with an inert carrier (e.g., pieces of polyurethane foam). Media were initially based on animal products (e.g., pig kidney) but were later improved by including various other ingredients (e.g., eggs, soy flour, peptone and yeast extract). The culture starts with the inoculation of the sterilised medium with bacteria followed by the nematodes. Nematodes are then harvested within 2–5 weeks by placing the foam onto sieves immersed in water. Only a few companies currently use this approach. A Chinese company Guangzhou Greenfine Biotechnology uses a solid culture method to produce several EPN species both for Chinese and international markets [57]. Other companies using this approach are Bionema (www.bionema.com), Andermatt Biocontrol AG (www.biocontrol.ch) and BioLogic USA (www.biologicco.com).

The *in vitro* liquid culture method is a complex process requiring medium development, understanding of the biology of the nematode-bacteria complex, the development of bioreactors and understanding and control of the process parameters. The process takes place in large bioreactors (up to 100.000 l). It is necessary to supply enough oxygen and prevent excessive shearing of the nematodes. Once the culture is completed, nematodes can be removed from the medium through centrifugation. This method is currently the most cost-effective [58], and thus the majority of EPN products result from liquid culture. Major producers using this method are BASF, Germany (www.agro.basf.com), E-Nema GmbH, Germany (www.e-nema.de), Koppert B.V., The Netherlands (www.koppert.com) etc.

3.2. Molluscoparasitic nematodes

In slug parasitic nematodes, there are two species that can be easily produced in a large scale, *P. hermaphrodita* and *A. appendiculatum*. The former is commercially produced as biocontrol agents while the later only for scientific purpose. *A. appendiculatum* can be easily produced on homogenised pig kidneys placed agar plates [62], but this nematode can be also mass produced in a solid Bedding medium [63] with a slight modification.

In vitro methods for mass production of *P. hermaphrodita* were developed in 1990s by Wilson [51]. Wilson showed that *P. hermaphrodita* can grow in a xenic culture in solid foam chip according to Bedding [63] and also in liquid cultures. Actually this species is the only commercially produced MPN. Technology used for producing of *P. hermaphrodita* is a modified method used for mass production of EPNs. The nematodes are produced in air-lift fermenters, up to 20,000 l or more in the balanced medium that allows yielding about 100,000 dauer juveniles in 1 ml of the medium. When the maximum yield is obtained, nematodes are concentrated by centrifuged. *P. hermaphrodita* is currently produced by BASF company (www.agro.basf.com) under the trademark Nemaslug©.

4. Formulation and application

4.1. Formulation of entomopathogenic nematodes

Entomopathogenic nematodes are always applied as infective juveniles and are mainly used for controlling the larval or pupal stages of insect pests in the soil or cryptic habitats. Under specific conditions, EPNs can successfully suppress also foliar pests [64].

EPNs have been classically applied in the form of aqueous suspension using sprayers, mist blowers, or irrigation systems. This approach turned out to have several limitations, mainly due to the sensitivity of the nematodes to desiccation and UV radiation [65]. For this reason, several alternatives improving formulation and application have been proposed and established.

4.1.1. Cadaver application

Insect cadaver application [66] has been proposed as a method enhancing EPN persistence. In this method, EPNs are applied in the infected insect host cadaver directly to the target site, and pest control is achieved by the infective juveniles that emerge from the host cadavers.

Insect cadaver application proved to be superior in EPN infectivity, survival, dispersal and pest control efficacy in some instances [67, 68]. EPN delivery can be further improved by formulating the infected hosts in coatings [69]. The cadaver application method has so far only been used commercially on a small scale relative to conventional methods [70], and it is especially useful for small- and medium-sized growers due to easier application and reduced storage costs [71].

Recently, the use of live insect hosts pre-infected with entomopathogenic nematodes against insect pests living in cryptic habitats was tested [72]. In this study, the release of the pre-infected lawn caterpillar, *Spodoptera cilium* (Lepidoptera: Noctuidae) against *S. cilium* in Bermudagrass arenas was as equally successful as standard aqueous application. The use of pre-infected *G.*

mellonella against the goat moth *Cossus cossus* (Lepidoptera: Cossidae) in chestnut (*Castanea sativa*) logs was much more efficient in comparison to the standard aqueous application. This novel approach thus showed an immense potential to control insect pests living in hard-to-reach cryptic habitats.

4.1.2. Capsules

Formulation of EPNs in polymer-based capsules can protect EPNs from desiccation and UV radiation and from biotic stressors such as their natural enemies. This approach was first used with *S. feltiae* and *H. bacteriophora* that were encapsulated in calcium alginate and fed to larvae of *Spodoptera exigua* [73]. In the following study, [74] tomato seeds were placed into the alginate matrix containing nematodes. When the seed germinated, the nematodes escaped from the capsule and could infect the host.

The higher efficiency can be further achieved by addition of another compatible pesticide [58]. Also, the recently proposed 'lure and kill' approach based on the application of the nematodes in capsules with insect attractant may reduce the number of nematodes necessary to control the insect pest as has been shown [75]. These authors developed alginate capsules containing EPNs and buried them in the rhizosphere of maize (*Zea mays*). The addition of attractants and feeding stimulants to the shell attracted the pest larvae as much as maize roots and in field trials, encapsulated *H. bacteriophora* nematodes were more effective in comparison to the nematodes applied in the aqueous suspension on the soil surface. Further studies improve capsule properties in order to increase EPN retainment within the capsules [76].

4.1.3. Shelf life

Besides aforementioned cadaver and gel formulations, EPNs are formulated in water-dispersible granules, nematode wool, gels, vermiculite, clay, peat, sponge, etc. The formulation, together with nematode species, strongly affects the shelf life of the EPN-based products. Actively moving nematodes are metabolically very active and use energy reserves soon [77]. Thus, they can remain alive and infective for 1–6 months under refrigeration ranges. EPNs with reduced mobility (formulations in gels) are still infective after up to 9 months of storage, whereas EPNs formulated in partial anhydrobiosis (formulations in water soluble powders) remain so for up to 1 year.

The root exudates were revealed to induce quiescence in EPNs that is reversible after placing the IJs in soil with high water content [78]. This approach could be used to prolong the shelf life of beneficial entomopathogenic nematodes (EPNs).

4.2. Formulation of molluscoparasitic nematodes

As was mentioned in the previous subchapter, the only commercial product based on MPNs (*P. hermaphrodita*) is Nemaslug© (BASF). Experiments with other nematodes species as bio-agents, for example, *A. appendiculatum* [42, 79] and some other rhabditids [80] were already done, but the results and the potential of these nematodes for the use in bio-control are still questionable and too far from practical impact.

General recommendation is to apply *P. hermaphrodita* on wet soil in the dose of 300,000 DJs/m^2 and water the soil immediately after application. The optimal application time is early evening when the soil temperature is about 15°C. Nematode efficacy can be increased by cultivation of soil just after application [81]. Nematodes are protected against UV radiation and drying. Nozzles and filters should have holes at least 1 mm wide, and the pressure should not exceed 5 bar. It is good to avoid application of *P. hermaphrodita* in the areas that were treated with some toxic chemicals, for example, pellets based on methiocarb used against noxious slugs. Combination with metaldehyde is safe for nematodes, because this compound does not affect them in concentration recommended for field application [82].

There are various strategies for the application. Common strategy is to apply the nematodes over the whole soil surface, and the alternative strategies are based on local applications. Slugs, *Deroceras reticulatum* and others, tend to avoid places treated with *P. hermaphrodita* [83]. Therefore, there were some ideas to apply the nematodes only around individual plants or in bands centred on plant rows. Unfortunately, the assumption of protecting plants using this approach with a lower amount of nematodes was not confirmed. There is no significant benefit associated with band or local application as opposed to uniform application [84]. The number of DJs decrease in time and the repellent effect to slugs subsides. The method of the reduction of the dose of nematodes but without lowering of the efficacy against slugs was published by Grewal et al. [85]. The principle is to apply nematodes in dose 0.6×10^6/m^2 only under artificial shelters that are used by slugs during day. This method provides almost the same effect as uniform application of 0.3×10^6/m^2. Highly effective can be repeated application of lower than recommended dose. In Brussels sprouts (*Brassica oleracea*) application of 50,000 DJs/m^2 is three times repeated in 1-month interval. It represents 50% reduction of the previously recommended single application, while the efficacy is almost the same as in case of using metaldehyde pellets [86].

P. hermaphrodita is applied in many plants in greenhouses, vegetables, ornamentals and in arable crops, for example, *Cymbidium* sp., lettuce, cabbage, Brussels sprouts, *Asparagus* sp., oilseed rape, wheat or sugarbeet and many other crops. The most common target pest are *Deroceras* sp. and *Arion* sp. Repeated uniform application on the soil surface is usual, but *P. hermaphrodita* can also be applied in the plastic tunnels or pots used in greenhouses. In arable crops, the nematodes have future especially in organic farms.

P. hermaphrodita is formulated in, for example, vermiculite [87] that slightly dehydrate and immobilise nematodes that can save energy more effectively in this state. Formulated nematodes are packed into polyethylene bags that allow exchange of air but retain water. The final product can be stored in a refrigerator for up to six months [47].

4.3. Genetic improvement

Genetic improvement has been an important contributor to the enormous advances in productivity that have been achieved over the past 50 years in plant and animal species that are of agricultural importance [88]. For entomopathogenic nematodes, main target characteristics are virulence, host range, heat and desiccation tolerance and shelf life. Glazer [89] summarised the four potential genetic-manipulation strategies: artificial selection, hybridisation, mutation and recombinant DNA techniques. Because it is unlikely that a transgenic EPN strain would meet public acceptance as a control agent [90], hybridisation and selective breeding are the most promising approaches to enhance EPN characteristics.

In a pioneer selection study performed with EPNs, the host-finding ability of *S. feltiae* was enhanced 20-fold to 27-fold after 13 selection rounds [91]. However, relaxation of the selection pressure produced a gradual decrease in host-finding. Similarly, Salame et al. [92] increased downward migration and infectivity of *S. feltiae.*

Many studies also attempted to enhance EPN tolerance to environmental stresses. Ehlers et al. [93] enhanced the low-temperature activity of *H. bacteriophora* by reducing the mean temperature at which the dauer juveniles (DJs) were active from 7.3 to 6.1°C during five selective breeding steps. Nimkingrat et al. [94] enhanced cold tolerance in *S. feltiae* by selecting and hybridizing the most cold-active strains. The cold tolerance was lost after few reproductive cycles under standard conditions, but was recovered after seven selection cycles with exposure to low temperatures.

Ehlers et al. [93] increased the mean tolerated temperature from 38.5 to 39.2°C. (The heritability for heat tolerance was 0.68 and for activity at low temperature 0.38). Salame et al. [92] bred a heterogeneous population of the EPN *Steinernema feltiae* for desiccation tolerance. A high survival rate (>85%) at 85% relative humidity for 72 h was obtained after 20 selection cycles. Mukuka et al. [95] searched for the most desiccation and heat tolerant strains of *H. bacteriophora*. In the following study [96], the authors crossed the most tolerant strains, and by subsequent selection they further increased desiccation and heat tolerance. Mean tolerated temperature of the most thermotolerant strain was 44°C after adaptation (vs. 38.2°C recorded for the commercial strain). The most desiccation tolerant strain had a mean tolerated water activity (aw-value) of 0.65 (vs. 0.951 in commercial strain).

Perry et al. [90] concluded that screening among natural populations for high tolerance to desiccation is a feasible approach and cross-breeding and genetic selection can further improve tolerance. However, there is a crucial question of the stability of selected traits. In *Heterorhabditis* nematodes, the trait stabilisation can be achieved by creation of inbred lines in liquid culture [97, 98].

According to Glazer [89] for EPNs, we lack markers to follow transfer or enhancement/ degradation of traits and to identify 'beneficial genes' that can be transferred between populations. Further fundamental research in the field of the genetic architecture of key traits, such as infectivity, stress tolerance and reproduction, is needed.

Thanks to recent advances in EPN and bacteria genomics [99] it will be possible to determine genes from the whole genome that are being expressed, in order to detect those that are involved in a particular process and target them through genetic engineering methods.

4.4. Safety

Entomopathogenic nematode-bacteria complexes are pathogens capable of invading and killing a large number of insects and even other arthropods, for example, spiders, ticks and millipedes [100]. It is thus necessary to establish the risk that these organisms applied for pest control pose to the environment and nontarget organisms.

Numerous studies have assessed the effect of these complexes on nontarget invertebrates, animals and humans and environment, and several conclusions can be drawn. The available data show that entomopathogenic nematode-bacteria complexes are generally safe to humans and animals, and their impact on nontarget insects and other invertebrates seems to be limited. An excellent review on this topic was given by Akhurst and Smith [101]. In this chapter, we shortly review the current knowledge and stress some recent findings.

4.4.1. *Safety to the environment*

Negative effect to the environment is likely to be much stronger if the introduced nematode establish in the target locality. Therefore, the establishment potential of the introduced beneficial nematodes represents a very important part of the risk assessment. The available information, however, is quite scarce and inconsistent. It has been shown that *H. bacteriophora* experimentally introduced to several fields in Germany persisted for a maximum of 2 years [102]. Exotic nematode *S. riobrave,* on the other hand, successfully established in the treated corn fields in USA [103]. Dillon et al. [104] reported the establishment of *S. feltiae* after application to forest clearcuts in Ireland, whereas *S. carpocapsae* and *H. bacteriophora* disappeared. Another example of the successful establishment is *S. scapterisci,* from Uruguay, was introduced in Florida, established in the target grassland areas, and even extended to other nonselected crops [105].

4.4.2. *Safety to nontarget invertebrates*

According to Bathon [106], the mortality of nontarget animals in the field may occur, but will be temporal, spatially restricted, affecting a part of the population, and its impact can be considered negligible. Piedra-Buena et al. [107] stated that the impact of EPNs in general on organisms considered 'non-target' is limited, with infections only occurring when these organisms are exposed to very high concentrations and under laboratory conditions.

Laboratory experiments have shown that EPNs can negatively affect a large number of invertebrates, including predatory insects [108], parasitoids [109, 110], Symphyla, Collembola, Arachnida, Crustacea, Diplopoda [111], terrestrial isopods, millipedes and Gastropods [112]. However, the field data generally show none or only a small reduction in field populations of nontarget species after applications of entomopathogenic nematodes [113, 114].

In a recent study, Dutka et al. [115] reported that *Bombus terrestris* is remarkably susceptible to two commercially available entomopathogenic nematode pest control products applied at the recommended field concentration. The authors imply that the fossorial habits of *B. terrestris,* and the overwintering of queens underground, may make this species uniquely vulnerable to biological pest control agents applied directly to the soil. However, it can be speculated that higher temperatures up to 30°C and a low relative humidity around 60% [116] within the bumblebee nest would not favour nematode infection and propagation.

4.4.3. *Safety to humans and animals*

Entomopathogenic nematode-bacteria complexes are generally considered safe to humans and animals. Many studies assessed the effect of EPNs on vertebrates. EPNs were applied orally, subcutaneously, peritoneally and intracerebrally to various vertebrates. In poikilotherms, the nematode application had usually no negative effect, with the exception of tadpoles, where nematode application caused mortality [117, 118]. However, the mortality was associated not with *Xenorhabdus* but with foreign bacteria entering the penetration holes made by the invading nematodes [101]. In homoioterms, no adverse effects have been recorded, with the exception of mice injected subcutaneously, where the nematodes caused the development of skin ulcers [119]. One case of possible human allergic response to EPNs was recorded in the person handling the concentrated nematode solutions during the harvesting, cleaning and storage stages of production [101].

The safety of bacterial symbionts has been tested by oral, intradermal, subcutaneous and intraperitoneal applications of the bacterial cells to various model vertebrates generally producing no adverse effect [120, 121]. There is, however, one exception, being *Photorhabdus asymbiotica*. Since 1989, some *Photorhabdus* strains have been identified as facultative human pathogens causing severe ulcerated skin lesions [122]. Ten years later, these clinical strains have been described as *P. asymbiotica* [123]. Mulley et al. [124] demonstrated that during a human infection, *P. asymbiotica* aggressively acquires amino acids, peptides and other nutrients from the human host, employing a so-called 'nutritional virulence' strategy. The authors further revealed that, interestingly, an insect Phenol-oxidase inhibitor Rhabduscin protects *P. asymbiotica* against the human complement pathway.

However, later studies identified also symbiotic strains of *P. asymbiotica* in association with *Heterorhabditis gerrardi* [125, 126] and *H. indica* [127], raising serious concerns about the safety of EPNs to humans.

European environmental risk assessment (ERA) excludes *Heterorhabditis indica* from the normal ERA exemption for EPNs, because of the rare association with this nematode of the symbiotic bacterium *Photorhabdus asymbiotica*. For this reason, there should be a precise identification of the symbiotic bacterium when *H. indica* is used for biocontrol [128].

Other commercially produced heterorhabditids, *H. bacteriophora* and *H. megidis*, have never been found in association with this bacterium and thus do not pose such a risk. Nevertheless, any contact between EPN-associated bacteria and human wounds should be avoided [129].

Very recently, Gengler et al. [130] have revealed the capacity of EPNs to act as an efficient reservoir ensuring exponential multiplication, maintenance and dissemination of the human pathogenic bacterium *Yersinia pseudotuberculosis*. The authors argue that if the similar relationship is between EPNs and *Y. pestis*, etiologic agent of plague, then it would enhance the understanding of long-term persistence of *Y. pestis* in plague endemic areas worldwide. Further research of this topic is necessary to determine any possible risk.

4.4.4. *Phasmarhabditis hermaphrodita*

The effect of commercial strain of *P. hermaphrodita* against many invertebrates has been tested in many studies. This organism is able to infect many slug and snail species, nontarget molluscs included. *Cepaea hortensis* and aquatic snail *Lymnaea stagnalis* are found susceptible to very high doses that several times exceed the recommended dose, whereas other aquatics mollusc, for example, *Physa fontinalis* are not [131–133]. Some other snails (*Succinaea putris, Pomatias elegans, Cepaea nemoralis* and others) can be infected with *P. hermaphrodita* but its effect on them is very low, if any. Negative effect on the earthworms *Lumbricus terrestris* and *Eisenia foetida* and others has never been found [134, 135], and no effect was found also against insects *Pterostichus melanarius, Zophobas morio* or *Galleria mellonella* [136, 137]. *P. hermaphrodita* is freely associated with many soil-dwelling bacteria [55, 138], and some of them, for example, *Stenotrophomonas maltophilia* [42] can be occasionally dangerous for human, especially those with a weakened immunity system.

4.5. Synergy with other biocontrol agents

Entomopathogenic and mollusc-parasitic nematodes are widely used in integrated and biological pest control systems. Entomopathogenic nematodes are relatively resistant to many pesticides in recommended dosage, except for some, for example, carbamates [82], and some authors reported synergy between EPNs and chemicals [139, 140]. But they are also influenced by many, especially soil dwelling, micro- and macro-organisms that can hardly suppress [141–143] or synergistically support them [144].

Synergy between entomopathogenic nematodes and other bio-agents are in great demand because this strategy can significantly reduce application rates and increase efficacy [145] that leads to higher economic profit. The great example of synergy between EPN *S. kraussei* (Nemasys L.) and insect-parasitic fungus *Metarhizium anisopliae* strain V275 was described [144]. Combination of a rates 1×10^{10} conidia and 250 000 IJs applied against overwintering larvae of black vine weevil *Otiorhynchus sulcatus* (Coleoptera: Curculionidae) resulted in 100% control, while the results in single applications were not so impressive. Similar results were obtained by Anbesse et al. [146] who tested synergistic effect of *H. bacteriophora* and *M. anisopliae* against barley chafer grub *Coptognathus curtipennis* (Coleoptera: Dynastidae) and Choo et al. [147] who reported synergy between *S. carpocapsae* and *Beauveria brongniartii* in control of oriental beetle *Exomala orientalis* (Coleoptera: Scarabeidae) grubs.

Synergistic effects were also found between EPNs and entomopathogenic bacteria *Bacillus thuringiensis* (Bt). Koppenhöfer and Kaya [148] reported additive and synergistic interaction between Bt and *S. glaseri* and *H. bacteriophora* that were applied against scarab grubs but also noted that these effects were not observed in case of *S. kushidai*. Similar reports about very low or absolutely no synergy between EPNs and other bio-agents, especially fungus and bacteria were published by many other authors [149, 150]. This inconsistence in results was explained by antagonism of nematodes symbiotic bacteria and other entomopathogens [151]. As stated in this study, bacteria *Photorhabdus luminiscens* is able to strongly suppress the growth and conidia production of *Beauveria bassiana, B. brongniarti* and *Paecilomyces fumosoroseus,* whereas other bacterial symbiont *Xenorhabdus poinari* does not. Shapiro-Illan et al. [152] provide that neutral or negative interactions among EPNs and other bio-agents are also dependent on the specific pathogens, hosts, application parameters and environmental conditions.

Interestingly, the use of combination of several EPN species has been shown to increase the efficacy against insect pests. There was a very strong synergy of *Steinernema weiseri* with *H. bacteriophora* or *S. glaseri* applied on *Curculio elephas* (Coleoptera: Curculionidae) a major pest of chestnut [153].

Reports of synergy of EPNs and arthropod bio-agents are slightly less frequent, maybe because of the ability of EPNs to infect many of these organisms, but despite this there are some successful combined applications that clearly show synergistic effect [154]. These authors reported positive effect of combined application of predatory mite *Hypoaspis aculeifer* and *H. bacteriophora* or *S. feltiae* against soil-dwelling stages of western flower thrips *Frankliniella occidentalis*. Positive effects of the combined applications of EPNs and arthropod bio-agents can be mostly expected when EPNs are used against soil-dwelling stages and arthropods against leaf-living stages of insect pests.

Expectably there was also synergism of EPNs in combination with GM plants [155]. Entomopathogenic nematodes are not negatively influenced by the GM plant and can infect all soil-dwelling stages of insect pest that survive or avoid the effect of GM plant (e.g., Bt-corn) that results in higher efficacy of biocontrol.

Even though there are some reports of antagonism among nematodes and other bio-agents, we can say that, in general, higher diversity of predators and similarly also pathogens leads to better control of many pests [156], thanks to the synergy of their effects on pest populations. Conservation of natural enemies may carry additional benefits for biological control.

Acknowledgements

This chapter was completed with the support of the Czech Ministry of Education, Youth and Sport in the programme Czech-Israeli Cooperative Scientific Research, project 8G15006.

Author details

Vladimír Půža*, Zdeněk Mráček and Jiří Nermuť

*Address all correspondence to: vpuza@seznam.cz

Biology Centre CAS, Institute of Entomology, České Budějovice, Czech Republic

References

[1] Hominick W.M. Biography. In. Gaugler R, editor. Entomopathogenic nematology. Wallingford, UK CAB International; 2002. p. 115-143.

[2] Mráček Z, Bečvář S, Kindlmann P, Jersáková J. Habitat preference for entomopathogenic nematodes, their insect hosts and new faunistic records for the Czech Republic. Biological Control. 2005;34:27-37.

[3] Stock PS, Pryor BM, Kaya HK. Distribution of entomopathogenic nematodes (Steinernematidae and Heterorhabditidae) in natural habitats in California. Biodiversity and Conservation. 1999;8:535-549.

[4] Malan AP, Knoetze R, Moore SD. Isolation and identification of entomopathogenic nematodes from citrus orchards in South Africa and their biocontrol potential against false codling moth. Journal of Invertebrate Pathology. 2011;108:115-125.

[5] Seenivasan N, Prabhu S, Makesh S, Sivakumar M. Natural occurrence of entomopathogenic nematode species (Rhabditida. Steinernematidae and Heterorhabditidae) in cotton fields of Tamil Nadu, India. Journal of Natural History. 2012;46:2829-2843.

[6] Barbosa-Negrisoli CR, Garcia MS, Dolinski C, Negrisoli AS, Jr, Bernardi D, dos Santos FJ. Survey of entomopathogenic nematodes (Rhabditida: Heterorhabditidae, Steinernematidae) in Rio Grande do Sul State, Brazil. Nematologia Brasileira. 2010;34:189-197.

[7] Hominick WM, Reid AP, Bohan DA, Briscoe BR. Entomopathogenic nematodes: biodiversity, geographical distribution and the convention on biological diversity. Biocontrol Science and Technology. 1996;6:317-331.

[8] Spiridonov SE, Lyons E, Wilson M. *Steinernema kraussei* (Rhabditida, Steinernematidae) from Island. Comparative Parasitology. 2004;71:215-220.

[9] Kalinkina SD, Spiridonov SE. First report of *Steinernema affine* (Bovien, 1937) Wouts, Mráček, Gerdin and Bedding, 1982 from the Russian Federation. Russian Journal of Nematology. 2015;23:153-154.

[10] Melo EL, Ortega CA, Susurluk A, Gaigl A, Bellotti AC. Native entomopathogenic nematodes (Rhabditida) in four departments of Colombia. Revista Colombiana de Entomologia. 2009;35:28-33.

[11] Parkman JP, Smart GC, (1996). Entomopathogenic nematodes, a case study: introduction of *Steinernema scapterisci* in Florida. Biocontrol of Science and Technology. 1996;6:413-419.

[12] Addis T, Mulawarman M, Waeyenberge L, Moens M, Viaene N, Ehlers RU. Identification and intraspecific variability of *Steinernema feltiae* strains from Cemoro Lawang village in Eastern Java, Indonesia. Russian Journal of Nematology. 2011;19:21-29.

[13] Akyazi F, Ansari MA, Ahmed BI, Crow WT, Mekete T. First record of entomopathogenic nematodes (Steinernematidae and Heterorhabditidae) from Nigerian soil and

their morphometrical and ribosomal DNA sequence analysis. Nematologia Mediterranea. 2012;40:95-100.

[14] Wang H, Luan JB, Dong H, Qian HT, Cong B. Natural occurrence of entomopathogenic nematodes in Liaoning (Northeast China). Journal of Asia-Pacific Entomology. 2014;17:399-406.

[15] Nguyen KB, Hunt DJ, Mracek Z. Steinernematidae: species descriptions. In: Nguyen KB, Hunt DJ, editors. Entomopathogenic nematodes: systematics, phylogeny and bacterial symbionts, nematology monographs and perspectives. Leiden-Boston: Brill; 2007. p. 121-609.

[16] Tarasco E, Clausi M, Rappazzo G, Panzavolta T, Curto G, Sorino R, Oreste M, Longo A, Leone D, Tiberi R, Vinciguerra MT, Triggiani O. Biodiversity of entomopathogenic nematodes in Italy. Journal of Helminthology. 2015;89:359-366.

[17] Mráček Z, Kindlmann P, Webster JM. *Steinernema affine* (Nematoda: Steinernematidae), a new record for North America and its distribution relative to other entomopathogenic nematodes in British Columbia. Nematology. 2005;7:495-501.

[18] Nguyen KB, Ambrós Ginarte CM, Leite LG, dos Santos JM, Harakava R. *Steinernema brazilense* n.sp. (Rhabditida: Steinernematidae) a new entomopathogenic nematode from Mato Grosso, Brazil. Journal of Invertebrate Pathology. 2010;103:8-20.

[19] Fischer-Le Saux M, Mauleon H, Constant P, Brunel B, Boemare N. PCR-ribotyping of *Xenorhabdus* and *Photorhabdus* isolates from the Caribbean region in relation to the taxonomy and geographic distribution of their nematode hosts. Applied and Environmental Microbiology. 1998;64:4246-4254.

[20] Hominick WM, Briscoe BR. Occurrence of entomopathogenic nematodes (Rhabditida: Steinernematidae and Heterorhabditidae) in British soils. Parasitology. 1990;100:295-302.

[21] Stuart RJ, Gaugler R. Patchiness in populations of entomopathogenic nematodes. Journal of Invertebrate Patholology. 1994;64:39-45.

[22] Sturhan D. Prevalence and habitat specificity of entomopathogenic nematodes in Germany. In: Proceedings of the Workshop COST 819; 16-2.9 May 1995; Italy. Luxembourg: European Commission; 1999. p. 123-132.

[23] Campos-Herrera R, Gomez-Ros JM, Escuer M, Cuadra L, Barrios L, Gutierrez C. Diversity, occurrence, and life characteristics of natural entomopathogenic nematode populations from La Rioja (Northern Spain) under different agricultural management and their relationships with soil factors. Soil Biology & Biochemistry. 2008;6:1474-1484.

[24] Ma JA, Chen SL, Zou YX, Li XH, Han RC, De Clercq P, Moens M. Natural occurrence of entomopathogenic nematodes in North China. Russian Journal of Nematology. 2010;18:117-126.

[25] Peters A. The natural host range of *Steinernema* and *Heterorhabditis* spp. and their impact on insect populations. Biocontrol Science and Technology. 1996;6:389-402.

[26] Mráček Z, Bečvář S. Insect aggregations and entomopathogenic nematode occurrence. Nematology. 2000;2:297-301.

[27] Mráček Z, Sturhan D. Epizootic of the entomopathogenic nematode, *Steinernema intermedium* (Poinar) in a nest of the bibionid fly, *Bibio marci* (L). Journal of Invertebrate Pathology. 2000;76:149-150.

[28] Alatorre-Rosas R, Kaya HK. Interspecific competition between entomopathogenic nematodes in the genera *Heterorhabditis* and *Steinernema* for an insect host in sand. Journal of Invertebrate Pathology. 1990;55:179-188.

[29] Koppenhofer AM, Kaya HK, Shanmugam S, Wood GL. Interspecific competition between steinernematid nematodes within an insect host. Journal of Invertebrate Pathology. 1995;66:99-103.

[30] Půža V, Mráček Z. Mixed infection of *Galleria mellonella* with two entomopathogenic nematodes (Nematoda: Rhabditida) species *Steinernema affine* benefits from the presence of *Steinernema kraussei*. Journal of Invertebrate Pathology. 2009;102:40-43.

[31] Půža V, Mráček Z. Mechanisms of coexistence of two sympatric entomopathogenic nematodes, *Steinernema affine* and *S. kraussei* (Nematoda: Steinernematidae), in a central European oak woodland soil. Applied Soil Ecology. 2010;45:65-70.

[32] Sturhan D, Mráček Z. Comparison of the *Galleria* baiting technique and a direct extraction method for recovering *Steinernema* infective-stage juveniles from soil. Folia Parasitologica. 2000;47:315-318.

[33] Campos-Herrera R, Johnson EG, El-Borai FE, Stuart RJ, Graham JH, Duncan LW. Long-term stability of entomopathogenic nematode spatial patterns in soil as measured by sentinel insects and real-time PCR assays. Annals of Applied Biology. 2011;158:55-68.

[34] Boemare NE, Akhurst RJ, Mourant RG. DNA relatedness between *Xenorhabdus spp.* (Enterobacteriaceae), symbiotic bacteria of entomopathogenic nematodes, and a proposal to transfer *Xenorhabdus luminescens* to a new genus, *Photorhabdus* gen. nov. International Journal of Systematic Bacteriology. 1993;43:249-255.

[35] Akhurst RJ, Boemare NE. Biology and taxonomy of *Xenorhabdus*. In: Gaugler R, Kaya HK, editors. Entomopathogenic nematodes in biological control. Boca Raton: CRC Press; 1990. p. 75-90.

[36] Akhurst RJ. Antibiotic activity of *Xenorhabdus* spp., bacteria symbiotically associated with insect pathogenic nematodes of the families Heterorhabditidae and Steinernematidae. Journal of General Microbiology. 1982;128:3061-3065.

[37] Chen G, Dunphy GB, Webster JM. Antimycotic activity of two *Xenorhadus* species and *Photorhabdus luminescens*, bacteria associated with the nematodes *Steinernema* species and *Heterorhabditis megidis*. Biological Control. 1994;4:339-348.

[38] Foltan P, Půža V. To complete their life cycle, pathogenic nematode–bacteria complexes deter scavengers from feeding on their host cadaver. Behavioural Processes. 2009;80:76-79.

[39] Rae RG, Robertson JF, Wilson MJ. The chemotactic response of *Phasmarhabditis hermaphrodita* (Nematoda: Rhabditida) to cues of *Deroceras reticulatum* (Mollusca: Gastropoda). Nematology. 2006;8:197-200.

[40] Nermuť J, Půža V, Mráček Z. The response of *Phasmarhabditis hermaphrodita* (Nematoda: Rhabditidae) and *Steinernema feltiae* (Nematoda: Steinernematidae) to different host-associated cues. Biological Control. 2012;61:201-206.

[41] Mengert H. Nematoden und Schnecken. Zeitschrift für Morphologie und Ökologie der Tiere. 1953;48:311-349.

[42] Nermuť J, Půža V, Mráček Z. The effect of different growing substrates on the development and quality of *Phasmarhabditis hermaphrodita* (Nematoda: Rhabditidae). Biocontrol Science and Technology. 2014;24:1026-1038.

[43] Nermuť J, Půža V, Mráček Z. Re-description of the slug-parasitic nematode *Alloionema appendiculatum* Schneider, 1859 (Rhabditida: Alloionematidae). Nematology. 2015;17:897-910.

[44] Ehlers RU, Shapiro-Ilan DI. Mass production. In: Grewal PS, Ehlers RU, Shapiro-Ilan DI, editors. Nematodes as biocontrol agents. Wallingford: CABI Publishing; 2005. p. 65-78.

[45] Abu Hatab M, Gaugler R, Ehlers RU. Influence of culture method on *Steinernema glaseri* lipids. Journal of Parasitology. 1998;84:215-221.

[46] Schneider A. Monographie der Nematoden. Berlin: Druck und Verlag von Georg Reimer; 1859.

[47] Morand S, Wilson MJ, Glen DM. Nematodes (Nematoda) parasitic in terrestrial gastropods. In: Barker GM, editor. Natural enemies of terrestrial molluscs. Wallingford: CABI Publishing; 2004. p. 525-557.

[48] Morand S. Cycle évolutif de *Nemhelix bakeri* Morand et Petter (Nematoda, Cosmocercidae) parasite de l'appareil génital de *Helix aspersa* (Gastropoda, Helicidae). Canadian Journal of Zoology. 1988;66:1796-1802.

[49] Morand S. Deux nouveaux nématodes Cosmocercidae parasites des escargots terrestres: *Cepaea nemoralis* L. et *Cepaea hortensis* Müller. Bulletin Museum National Histoire Naturelle, Paris. 1989;11:563-570.

[50] Mráček Z,Weiser J. Parazitické hlístice hmyzu. Praha: Academia; 1988.

[51] Wilson MJ, Glen DM, George SK. The rhabditid nematode *Phasmarhabditis hermaphrodita* as a potential biological-control agent for slugs. Biocontrol Science and Technology. 1993;3:503-511.

[52] Rae RG, Robertson JF, Wilson MJ. Chemoattraction and host preference of the gastropod parasitic nematode *Phasmarhabditis hermaphrodita*. Journal of Parasitology. 2009;95:517-526.

[53] MacMillan K, Haukeland S, Rae R, Young I, Crawford J, Hapca S, Wilson MJ. Dispersal patterns and behaviour of the nematode *Phasmarhabditis hermaphrodita* in mineral soils and organic media. Soil Biology & Biochemistry. 2009;41:1483-1490.

[54] Wilson MJ, Glen DM, George SK, Pearce JD. Selection of a bacterium for the mass-production of *Phasmarhabditis hermaphrodita* (Nematoda: Rhabditidae) as a biocontrol agent for slugs. Fundamental and Applied Nematology. 1995;18:419-425.

[55] Rae RG, Tourna M, Wilson MJ. The slug parasitic nematode *Phasmarhabditis hermaphrodita* associates with complex and variable bacterial assemblages that do not affect its virulence. Journal of Invertebrate Pathology. 2010;104:222-226.

[56] Wilson MJ, Glen DM, Pearce JD, Rodgers PB. Monoxenic culture of the slug parasite *Phasmarhabditis hermaphrodita* (Nematoda: Rhabditidae) with different bacteria in liquid and solid-phase. Fundamental and Applied Nematology. 1995;18:159-166.

[57] Shapiro-Ilan DI, Han R, Qiu X. Production of entomopathogenic nematodes. In: Morales-Ramos J, Rojas G, Shapiro-Ilan D, editors. Mass production of beneficial organisms: invertebrates and entomopathogens. Amsterdam: Academic; 2014. p. 321-356.

[58] Shapiro-Ilan DI, Han R, Dolinksi C. Entomopathogenic nematode production and application technology. Journal of Nematology. 2012;44:206-217

[59] Glaser RW. Continued culture of a nematode parasitic in the Japanese beetle. Journal of Experimental Zoology. 1940:84:1

[60] Stoll NR. Axenic cultivation of the parasitic nematode, *Neoaplectana glaseri,* in a fluid medium containing raw liver extract. Journal of Parasitology. 1952;39:422-444.

[61] Lunau S, Stoessel S, Schmidt-Peisker AJ, Ehlers RU. Establishment of monoxenic inocula for scaling up in vitro cultures of the entomopathogenic nematodes *Steinernema* spp. and *Heterorhabditis* spp. Nematologica. 1993;39:385-399.

[62] Nermuť J, Půža V, Mráček Z. Some observations on morphology and ecology of mollusc-parasitic nematode *Alloionema appendiculatum*. In: Proceedings of the 47th annual meeting of the Society for Invertebrate Pathology; 3-7 August 2014; Mainz, Germany: 2014. p. 125-126.

[63] Bedding RA. Large scale production, storage and transport of the insect parasitic nematodes *Neoaplectana* spp. and *Heterorhabditis* spp. Annals of Applied Biology. 1984;104:117-120.

[64] Shapiro-Ilan DI, Gouge DH, Piggott SJ, Fife JP. Application technology and environmental considerations for use of entomopathogenic nematodes in biological control. Biological Control. 2006;38:124-133.

[65] Gaugler R, Wilson M, Shearer P. Field release and environmental fate of a transgenic entomopathogenic nematode. Biological Control. 1997;9:75-80.

[66] Shapiro-Ilan DI, Lewis EE, Behle RW, McGuire MR. Formulation of entomopathogenic nematode-infected cadavers. Journal of Invertebrate Pathology. 2001;78:17-23.

[67] [68] Perez EE, Lewis EE, Shapiro-Ilan DI. Impact of host cadaver on survival and infectivity of entomopathogenic nematodes (Rhabditida: Steinernematidae and Heterorhabditidae) under desiccating conditions. Journal of Invertebrate Pathology. 2003;82:111-118.

[68] Shapiro-Ilan DI, Lewis EE, Tedders WL, Son Y. Superior efficacy observed in entomopathogenic nematodes applied in infected-host cadavers compared with application in aqueous suspension. Journal of Invertebrate Pathology. 2003;83:270-272.

[69] Shapiro-Ilan DI, Morales-Ramos JA, Rojas MG, Tedders WL. Effects of a novel entomopathogenic nematode-infected host formulation on cadaver integrity, nematode yield, and suppression of *Diaprepes abbreviatus* and *Aethina tumida* under controlled conditions. Journal of Invertebrate Pathology. 2010;103:103-108.

[70] Lacey LA, Grzywacz D, Shapiro-Ilan DI, Frutos R, Brownbridge M, Goettel MS. Insect pathogens as biological control agents: back to the future. Journal of Invertebrate Pathology. 2015;132:1-41.

[71] Dolinski C, Shapiro-Ilan D, Lewis EE. Insect cadaver applications: pros and cons. In: Campos-Herrera R, editor. Nematode pathogenesis of insects and other pests. Springer International Publishing; 2015. p. 207-229. DOI:10.1007/978-3-319-18266-7_8

[72] Gumus A, Karagoz M, Shapiro-Ilan D, Hazir S. A novel approach to biocontrol: release of live insect hosts pre-infected with entomopathogenic nematodes. Journal of Invertebrate Pathology. 2015;130:56-60.

[73] Kaya HK, Nelsen CE. Encapsulation of steinernematid and heterorhabditid nematodes with calcium alginate: a new approach for insect control and other applications. Environmental Entomology. 1985;14:572-574.

[74] Kaya HK, Mannion CM, Burlando TM, Nelsen CE. Escape of *Steinernema feltiae* from alginate capsules containing tomato seeds. Journal of Nematology. 1987;19:287.

[75] Hiltpold I, Hibbard BE, French BW, Turlings TC. Capsules containing entomopathogenic nematodes as a Trojan horse approach to control the western corn rootworm. Plant and Soil. 2012;358:11-25.

[76] Kim J, Jaffuel G, Turlings TC. Enhanced alginate capsule properties as a formulation of entomopathogenic nematodes. BioControl. 2015;60:527-535.

[77] Lewis EE. Handling, transport, and storage of insecticidal nematodes. In: Proceedings of the workshop optimal use of insecticidal nematodes in pest management; New Jersey: Rutgers University; 1999. p. 25-30.

[78] Jaffuel G, Hiltpold I, Turlings, TC. Highly potent extracts from pea (*Pisum sativum*) and maize (*Zea mays*) roots can be used to induce quiescence in entomopathogenic nematodes. Journal of Chemical Ecology. 2015;41:793-800.

[79] Laznik Z, Ross JL, Trdan S. Massive occurrence and identification of the nematode *Alloionema appendiculatum* Schneider (Rhabditida: Alloionematidae) found in Arionidae slugs in Slovenia. Acta Agriculturae Slovenica. 2010;95:43-49.

[80] Charwat SM, Davies KA. Laboratory screening of nematodes isolated from South Australia for potential as biocontrol agents of helicid snails. Journal of Invertebrate Pathology. 1999;74:55-61.

[81] Hass B, Glen DM, Brain P, Hughes LA. Targeting biocontrol with the slug-parasitic nematode *Phasmarhabditis hermaphrodita* in slug feeding areas: a model study. Biocontrol Science and Technology. 1999;9:587-598.

[82] Nermuť J, Mráček Z. The influence of pesticides on the viability and infectivity of entomopathogenic nematodes (Nematoda: Steinernematidae). Russian Journal of Nematology. 2010;18:141-148.

[83] Wilson MJ, Hughes LA, Jefferie D, Glen DM. Slugs (*Deroceras reticulatum* and *Arion ater* agg.) avoid soil treated with the rhabditid nematode *Phasmarhabditis hermaphrodita*. Biological Control. 1999;16:170-176.

[84] Hass B, Hughe LA, Glen DM. Overall versus band application of the nematode *Phasmarhabditis hermaphrodita* with and without incorporation into soil, for biological control of slugs in winter wheat. Biocontrol Science and Technology. 1999;9:579-586.

[85] Grewal PS, Grewal SK, Taylor RAJ, Hammond RB. Application of molluscicidal nematodes to slug shelters: a novel approach to economic biological control of slugs. Biological Control. 2001;22:72-80.

[86] Ester A, van Rozen K, Molendijk LPG. Field experiments using the rhabditid nematode *Phasmarhabditis hermaphrodita* or salt as control measures against slugs in green asparagus. Crop Protection. 2003;22:689-695.

[87] Wilson MJ, Grewal PS. Biology, production and formulation of slug-parasitic nematodes. In: Grewal PS, Ehlers RU, Shapiro-Ilan DI, editors. Nematodes as biocontrol agents. Wallingford: CABI Publishing; 2005. p. 421-429.

[88] Dekkers JC, Hospital F. The use of molecular genetics in the improvement of agricultural populations. Nature Reviews Genetics. 2002;3:22-32.

[89] Glazer I. Improvement of entomopathogenic nematodes: a genetic approach. In: Campos-Herrera R, editor. Nematode pathogenesis of insects and other pests. Springer International Publishing; 2015. p. 29-55. DOI: 10.1007/978-3-319-18266-7_2

[90] Perry RN, Ehlers RU, Glazer I. A realistic appraisal of methods to enhance desiccation tolerance of entomopathogenic nematodes. Journal of Nematology. 2012;44:185-190.

[91] Gaugler R, Campbell JF, Mcguire TR. Selection for host finding in *Steinernema feltiae*. Journal of Invertebrate Pathology. 1989;54:363-372.

[92] Salame L, Glazer I, Chubinishvilli MT, Chkhubianishvili T. Genetic improvement of the desiccation tolerance and host-seeking ability of the entomopathogenic nematode *Steinernema feltiae*. Phytoparasitica. 2010;38:359-368.

[93] Ehlers RU, Oestergaard J, Hollmer S, Wingen M, Strauch O. Genetic selection for heat tolerance and low temperature activity of the entomopathogenic nematode–bacterium complex *Heterorhabditis bacteriophora–Photorhabdus luminescens*. BioControl. 2005;50:699-716.

[94] Nimkingrat P, Khanam S, Strauch O, Ehlers RU. Hybridisation and selective breeding for improvement of low temperature activity of the entomopathogenic nematode *Steinernema feltiae*. BioControl. 2013;58:417-426.

[95] Mukuka J, Strauch O, Ehlers RU. Variability in desiccation tolerance among different strains of the entomopathogenic nematode *Heterorhabditis bacteriophora*. Nematology. 2010;12:711-720.

[96] Mukuka J, Strauch O, Hoppe C, Ehlers RU. Improvement of heat and desiccation tolerance in *Heterorhabditis bacteriophora* through cross-breeding of tolerant strains and successive genetic selection. BioControl. 2010;55:511-521.

[97] Anbesse S, Sumaya NH, Dörfle, AV, Strauch O, Ehlers RU. Selective breeding for desiccation tolerance in liquid culture provides genetically stable inbred lines of the entomopathogenic nematode *Heterorhabditis bacteriophora*. Applied Microbiology and Biotechnology. 2013;97:731-739.

[98] Anbesse S, Sumaya NH, Dörfler AV, Strauch O, Ehlers RU. Stabilisation of heat tolerance traits in *Heterorhabditis bacteriophora* through selective breeding and creation of inbred lines in liquid culture. BioControl. 2013;58:85-93.

[99] Bai X, Adams BJ, Ciche TA, Clifton S, Gaugler R, Kim KS, Spieth J, Sternberg PW, Wilson RK, Grewal PS. A lover and a fighter: the genome sequence of an entomopathogenic nematode *Heterorhabditis bacteriophora*. PLoS One. 2013;8.

[100] Poinar GO Jr. Nematodes for biological control of insects. Florida: CRC Press Inc; 1979. 249 p.

[101] Akhurst R, Smith K, Regulation and safety. In: Gaugler R, editor. Entomopathogenic nematology. New York: CABI Publishing; 2002. p. 311-332.

[102] Susurluk A, Ehlers RU. Field persistence of the entomopathogenic nematode *Heterorhabditis bacteriophora* in different crops. BioControl. 2008;53:627-641.

[103] Millar LC, Barbercheck ME. Interaction between endemic and introduced entomopathogenic nematodes in conventional-till and no-till corn. Biological Control. 2001;22:235-245.

[104] Dillon AB, Rolston AN, Meade CV, Downes MJ, Griffin CT. Establishment, persistence, and introgression of entomopathogenic nematodes in a forest ecosystem. Ecological Applications. 2008;18:735-747.

[105] Campos-Herrera R, El-Borai FE, Stuart RJ, Graham JH, Duncan LW. Entomopathogenic nematodes, phoretic *Paenibacillus* spp., and the use of real time quantitative PCR to explore soil food webs in Florida citrus groves. Journal of Invertebrate Pathology. 2011;108:30-39.

[106] Bathon H. Impact of entomopathogenic nematodes on non-target hosts. Biocontrol Science and Technology. 1996;6:421-434

[107] Piedra-Buena A, López-Cepero J, Campos-Herrera R. Entomopathogenic nematode production and application: regulation, ecological impact and non-target effects. In: Campos-Herrera R, editor. Nematode pathogenesis of insects and other pests. Springer International Publishing; 2015. p. 255-282. DOI:10.1007/978-3-319-18266-7_10.

[108] Georgis R, Kaya HK, Gaugler R. Effect of steinernematid and heterorhabditid nematodes (Rhabditida: Steinernematidae and Heterorhabditidae) on nontarget arthropods. Environmental Entomology. 1991;20:815-822.

[109] Kaya HK. Infectivity of *Neoaplectana carpocapsae* and *Heterorhabditis heliothidis* to pupae of the parasite *Apanteles militaris*. Journal of Nematology. 1978;10:241-244.

[110] Kaya HK. Effect of the entomogenous nematode *Neoaplectana carpocapsae* on the tachinid parasite *Compsilura concinnata* (Diptera: Tachinidae). Journal of Nematology. 1984;16:9-13.

[111] Poinar GO, Jr. Non-insect hosts for the entomogenous rhabditoid nematodes *Neoaplectana* (Steinernematidae) and *Heterorhabditis* (Heterorhabditidae). Revue de Nématologie. 1989;12:423-428.

[112] Jaworska M. Infection of the terrestrial isopod Porcellio scaber Larr. and millipede *Blaniulus guttulatus* Bosc. with entomopathogenic nematodes (Nematoda: Rhabditidae) in laboratory conditions. Folia Horticulturae. 1991;3:115-120.

[113] Buck M, Bathon H. Effects of a field application of entomopathogenic nematodes (*Heterorhabditis* sp.) on the nontarget fauna, part 2: Diptera. Anzeiger für Schädlingskunde Pflanzenschutz Umweltschutz. 1993;66:84-88.

[114] Ropek D, Jaworska M. Effect of an entomopathogenic nematode, *Steinernema carpocapsae* Weiser (Nematoda: Steinernematidae), on a carabid beetles in field trials with annual legumes. Anzeiger für Schädlingskunde Pflanzenschutz Umweltschutz. 1994;67:97-100.

[115] Dutka A, McNulty A, Williamson SM. A new threat to bees? Entomopathogenic nematodes used in biological pest control cause rapid mortality in *Bombus terrestris*. PeerJ. 2015;3:e1413. DOI 10.7717/peerj.1413

[116] Weidenmüller A, Kleineidam C, Tautz J. Collective control of nest climate parameters in bumblebee colonies. Animal Behaviour. 2002;63:1065-1071.

[117] Kermarrec A, Mauléon H. Potential noxiousness of the entomogenous nematode *Neoaplectana carpocapsae* Weiser to the Antillan toad *Bufo marinus* L. Mededelingen Faculteit Landbouwkundige University of Gent. 1985;50:831-838.

[118] Kermarrec A, Mauléon H, Sirjusingh C, Baud L. Experimental studies on the sensitivity of tropical vertebrates (toads, frogs and lizards) to different species of entomoparasitic nematodes of the genera *Heterorhabditis* and *Steinernema*. In: Pavis C. Kermarrec A. editors. Rencontres Caraïbes en Lutte Biologique. Paris: Institut National de la Recherche Agronomique; 1991. p. 193-204.

[119] Kobayashi M, Okano H, Kirihara S. The toxicity of steinernematid and heterorhabditid nematodes to male mice. In: Ishibashi N. editor. Recent advances in biological control of insect pests by entomogenous nematodes in Japan. Japan: Saga University; 1987. p. 153-157.

[120] Poinar GO, Thomas GM, Presser SB, Hardy JL. Inoculation of entomogenous nematodes, *Neoaplectana* and *Heterorhabditis* and their associated bacteria, *Xenorhabdus* spp., into chicks and mice. Environmental Entomology. 1982;11:137-138.

[121] Obendorf DL, Peel B, Akhurst RJ, Miller LA. Non-susceptibility of mammals to the entomopathogenic bacterium *Xenorhabdus nematophilus*. Environmental Entomology. 1983;12:368-370.

[122] Farmer JJ, Jorgensen JH, Grimont PAD, Akhurst RJ, Poinar GO, Ageron E, Pierce GV, Smith JA, Carter GP, Wilson KL. *Xenorhabdus luminescens* (DNA Hybridization Group 5) from human clinical specimens, Journal of Clinical Microbiology. 1989;27:1594-1600.

[123] Fischer-Le Saux M, Viallard V, Brunel B, Normand P, Boemare NE. Polyphasic classification of the genus *Photorhabdus* and proposal of new taxa: *P. luminescens* subsp.

luminescens subsp. nov., *P. luminescens* subsp. *akhurstii* subsp. nov., *P. luminescens* subsp. *laumondii* subsp. nov., *P. temperata* sp. nov., *P. temperata* subsp. *temperata* subsp. nov. and *P. asymbiotica* sp. nov. International Journal of Systematic and Evolutionary Microbiology. 1999;49:1645-1656.

[124] Mulley G, Beeton ML, Wilkinson P, Vlisidou I, Ockendon-Powell N, Hapeshi A, Waterfield NR. From insect to man: *Photorhabdus* sheds light on the emergence of human pathogenicity. PLoS One. 2015;10(12):e0144937.

[125] Gerrard JG, Joyce SA, Clarke DJ, Ffrench-Constant RH, Nimmo GR, Looke DFM, Feil E, Pearce L, Waterfield N. Nematode symbiont for *Photorhabdus asymbiotica*. Emerging Infectious Diseases. 2006;12:1562-1564

[126] Plichta KL, Joyce SA, Clarke D, Waterfield N, Stock SP. *Heterorhabditis gerrardi* n. sp. (Nematoda: Heterorhabditidae): the hidden host of *Photorhabdus asymbiotica* (Enterobacteriaceae: gamma-Proteobacteria). Journal of Helminthology. 2009;16:1-12

[127] Kuwata R, Yoshiga T, Yoshida M, Kondo E. Mutualistic association of *Photorhabdus asymbiotica* with Japanese heterorhabditid entomopathogenic nematodes. Microbes and Infection. 2008;10:734-741.

[128] Bale J. Harmonization of regulations for invertebrate biocontrol agents in Europe: progress, problems and solutions. Journal of Applied Entomology. 2011;135:503-513.

[129] Grewal PS. Entomopathogenic nematodes as tools in integrated pest management. In: Abrol DP, Shankar U, editors. Integrated pest management: principles and practice. Wallingford: CABI; 2012. p. 162-236.

[130] Gengler S, Laudisoit A, Batoko H, Wattiau P. Long-term persistence of *Yersinia pseudotuberculosis* in entomopathogenic nematodes. PLoS One. 2015;10:e0116818. doi: 10.1371/journal. pone.0116818

[131] Wilson MJ, Hughes LA, Hamacher GM, Glen DM. Effects of *Phasmarhabditis hermaphrodita* on non-target molluscs. Pest Management Science. 2000;56:711-716.

[132] Coupland JB. Susceptibility of helicid snails to isolates of the nematode *Phasmarhabditis hermaphrodita* from Southern France. Journal of Invertebrate Pathology. 1995;66: 207-208.

[133] Morley NJ, Morritt D. The effects of the slug biological control agent, *Phasmarhabditis hermaphrodita* (Nematoda), on non-target aquatic molluscs. Journal of Invertebrate Pathology. 2006;92:112-114.

[134] Grewal SK, Grewal PS. Survival of earthworms exposed to the slug-parasitic nematode *Phasmarhabditis hermaphrodita*. Journal of Invertebrate Pathology. 2003;82:72-74.

[135] De Nardo EAB, Sindermann AB, Grewal SK, Grewal PS. Non-susceptibility of earthworm *Eisenia fetida* to the rhabditid nematode *Phasmarhabditis hermaphrodita*, a biocontrol agent of slugs. Biocontrol Science and Technology. 2004;14:93-98.

[136] Wilson MJ, Glen DM, Hughes LA, Pearce JD, Rodgers PB. Laboratory tests of the potential of entomopathogenic nematodes for the control of field slugs (*Deroceras reticulatum*). Journal of Invertebrate Pathology. 1994;64:182-187.

[137] Rae R, Verdun C, Grewal P, Robertson JF, Wilson MJ. Biological control of terrestrial molluscs using *Phasmarhabditis hermaphrodita*: progress and prospects. Pest Management Science. 2007;63:1153-1164.

[138] Wilson MJ, Glen DM, Pearce JD, Rodgers PB. Effects of different bacterial isolates on growth and pathogenicity of the slug-parasitic nematode *Phasmarhabditis hermaphrodita*. Proceedings Brighton Crop Protection Conference, Pests and Diseases. 1994;3:1055-1060.

[139] Koppenhöfer AM, Fuzy EM. Comparison of neonicotinoid insecticides as synergists for entomopathogenic nematodes. Biological Control. 2002;24:90-97.

[140] Koppenhöfer AM, Fuzy EM 2. Effect of the anthranilic diamide insecticide, chlorantraniliprole, on *Heterorhabditis bacteriophora* (Rhabditida: Heterorhabditidae) efficacy against white grubs (Coleoptera: Scarabaeidae). Biological Control. 2008;45:93-102.

[141] Strong D. Populations of entomopathogenic nematodes in food webs. In: Gaugler R, editor. Entomopathogenic nematology. New York: CABI Publishing; 2002. p. 225-240.

[142] Wilson M, Gaugler R. Factors limiting short term persistence of entomopathogenic nematodes. Journal of Applied Entomology. 2004;128:250-253.

[143] Nermuť J. The persistence of *Phasmarhabditis hermaphrodita* (Rhabditida: Rhabditidae) in different substrates. Russian Journal of Nematology. 2012;20:61-64.

[144] Ansari MA, Shah FA, Butt TM. The entomopathogenic nematode *Steinernema kraussei* and *Metarhizium anisopliae* work synergistically in controlling overwintering larvae of the black vine weevil, *Otiorhynchus sulcatus*, in strawberry growbags. Biocontrol Science and Technology. 2010;20:99-105.

[145] Ansari MA, Shah FA, Tirry L, Moens M. Field trials against Hoplia philanthus (Coleoptera: Scarabaeidae) with a combination of an entomopathogenic nematode and the fungus *Metarhizium anisopliae* CLO 53. Biological Control. 2006;39:453-459.

[146] Anbesse SA, Adge BJ, Gebru WM. Laboratory screening for virulent entomopathogenic nematodes (*Heterorhabditis bacteriophora* and *Steinernema yirgalemense*) and fungi (*Metarhizium anisopliae* and *Beauveria bassiana*) and assessment of possible synergistic effects of combined use against grubs of the barley chafer *Coptognathus curtipennis*. Nematology. 2008;10:701-709.

[147] Choo HY, Kaya HK, Huh J, Lee DW, Kim HH, Lee SM, Choo YM. Entomopathogenic nematodes (*Steinernema* spp. and *Heterorhabditis bacteriophora*) and a fungus *Beauveria brongniartii* for biological control of the white grubs, *Ectinohoplia rufipes* and *Exomala orientalis*, in Korean golf courses. BioControl. 2002;47:177-192.

[148] Koppenhöfer AM, Kaya HK. Additive and synergistic interaction between entomopathogenic nematodes and *Bacillus thuringiensis* for scarab grub control. Biological Control. 1997;8:131-137.

[149] Wu S, Youngman RR, Kok LT, Laub CA, Pfeiffer DG. Interaction between entomopathogenic nematodes and entomopathogenic fungi applied to third instar southern masked chafer white grubs, *Cyclocephala lurida* (Coleoptera: Scarabaeidae), under laboratory and greenhouse conditions. Biological Control. 2014;76:65-73.

[150] Saito T, Brownbridge M. Compatibility of soil-dwelling predators and microbial agents and their efficacy in controlling soil-dwelling stages of western flower thrips *Frankliniella occidentalis*. Biological Control. 2016;92:92-100.

[151] Ansari MA, Tiry L, Moens M. Antagonism between entomopathogenic fungi and bacterial symbionts of entomopathogenic nematodes. BioControl. 2005;50:465-475.

[152] Shapiro-Ilan DI, Jackson M, Reilly CC, Hotchkiss MW. Effects of combining an entomopathogenic fungi or bacterium with entomopathogenic nematodes on mortality of *Curculio caryae* (Coleoptera: Curculionidae). Biological Control. 2004;30:119-126.

[153] Demir S, Karagoz M, Hazir S, Kaya HK. Evaluation of entomopathogenic nematodes and their combined application against *Curculio elephas* and *Polyphylla fullo* larvae. Journal of Pest Science. 2014;88:163-170.

[154] Premachandra WTSD, Borgemeister C, Berndt O, Ehlers RU, Poehling HM. Combined releases of entomopathogenic nematodes and the predatory mite *Hypoaspis aculeifer* to control soil-dwelling stages of the western flower thrips *Frankliniella occidentalis*. Biocontrol. 2003;48:529-541.

[155] Gassmann AL, Stock SP, Sisterson MS, Carrière Y, Tabashnik BE. Synergism between entomopathogenic nematodes and *Bacillus thuringiensis* crops: integrating biological control and resistance management. Journal of Applied Ecology. 2008;45:957-966.

[156] Jabbour R, Crowder DW, Aultman EA, Snyder WE. Entomopathogen biodiversity increases host mortality. Biological Control. 2011;59:277-283.

Chapter 5

Synthesis and Application of Pheromones for Integrated Pest Management in Vietnam

Chi-Hien Dang, Cong-Hao Nguyen, Chan Im and Thanh-Danh Nguyen

Additional information is available at the end of the chapter

http://dx.doi.org/10.5772/63768

Abstract

The negative impacts of conventional pesticides on health, environment, and organisms have involved strong development of integrated pest management (IPM) strategies. The use of insect pheromones becomes an effectively alternative selection in agricultural and forest pest control. Pheromone researches in Vietnam started in the last few decades and in addition to technical factors, recent achievements in the Vietnamese agriculture have an important direct link to the pheromone developments. In this chapter, we review the pheromone researches related to synthesis and field trials of several especial insect pheromones, in which Vietnamese scientists have mainly participated or collaborated with foreign research groups. First, we will discuss an overview of popular insect pheromones in Vietnam, a lot of species of which are also found around the world, as an important reference for scientists who would have especial consideration in this field. Further, synthetic routes of pheromones are summarized with various structures including chiral, racemic, mono- and poly-olefinic pheromones where some schemes have become standard methodologies for synthesis of similar structural compounds. Finally, field evaluations of the pheromones of numerous species are discussed in detail.

Keywords: insect attractant, pheromone trap, synthesis, pest control, integrated pest management, field application

1. Introduction

Nowadays, numbers of pests are becoming increasingly resistant due to the conventional pesticides which cause damage to useful parasites and imbalance in the ecosystem, creating environmental pollution and adverse effect on the economy [1]. This leads to the concept of IPM

being rapidly developed to solve the problems of pesticide use. IPM allows safer insect control and poses the least risks while maximizing benefits and reducing costs. Although many methods have been employed, the use of pheromones obviously is one of the most effective approaches in pest control which can be achieved by mass trapping and killing the harmful pests selectively [2]. Moreover, the use of pheromone traps minimizes risks to human health and reduces destruction of the living environment.

Because of the tropical climate, Vietnam is generally favourable for many typical pests with rapid breeding which cause damage to the crops and forest trees throughout the year [3]. Since Nguyen group's first report in early 1980s [4], the demand of pheromone for IPM in Vietnam is steadily growing and that generally leads to a strong development of synthesis and field trial of pheromones. Success of the Vietnamese agriculture in the recent years has had important contribution from use of this technique. At the same time, the numerous reports on pheromones in Vietnam from other groups, such as groups of Can Tho University, have been increasing rapidly during the last decades.

Herein, our aim is to review synthesis and field trials of insect pheromones from research groups in Vietnam. The chapter consists of three main parts including introduction of pheromones of popular insects in Vietnam, an overview of synthetic methodology presented in structurally typical order, and finally field trials of several important insect pheromones.

2. Popular insect pheromones in Vietnam

Due to the big difference of weather conditions between the regions of Vietnam, for instance the South of Vietnam having no winter season but the North having four clearly identifiable seasons, numerous insect species have been found in Vietnam [3]. Known pheromones of popular species in the country are summarized in **Table 1** where only major component of pheromones identified at ratios of more than 65% in their mixture is listed with a respective reference. Species specificity is commonly achieved by the use of blends of several pheromone components. The data reveal that Lepidoptera is the best studied insect order related to pheromone, with data available for about 27 species. The pheromones of this order usually consist of alcohols, acetates, or aldehydes of long chain containing double bonds while the other orders mostly possess pheromone molecules bearing chiral carbons.

Order, family and species	Common name	Major component of pheromone*	Ref.
Lepidoptera			
Crambidae			
Cnaphalocrocis medinalis G.	Rice leaf folder	Z13-18:Ac or Z13-18:Ald	[5, 6]
Chilo suppressalis W.	Rice stem borer	Z11-16:Ald	[7]
Hellula undalis F.	Cabbage webworm	*E*11*E*13-16:Ald	[8]
Crocidolomia binotalis Z.	Cabbage head caterpillar	Z11-16:Ac	[9]

Order, family and species	Common name	Major component of pheromone*	Ref.
Conogethes punctiferalis G.	Yellow peach moth	*E*10-16:Ald	[10]
Scirpophaga nivella F.	Sugarcane top borer	*E*11-16:Ald	[11]
Proceras venosatus W.	Striped sugarcane borer	*Z*13-18:OH	[12]
Noctuidae			
Sesamia inferens W.	Pink stem borer	*Z*11-16:Ac	[13]
Helicoverpa armigera H.	Cotton bollworm	*Z*11-16:Ald	[14]
Spodoptera exigua H.	Beet armyworm	*Z*9*E*12-14:Ac	[15]
Spodoptera litura F.	Oriental leafworm	*Z*9*E*11-14:Ac	[16]
Chrysodeixis eriosoma D.	Green garden looper	*Z*7-12:Ac	[17, 18]
Ctenoplusia albostriata B. & G.	Eastern streaked plusia	*Z*7-12:Ac	[18]
Argyrogramma signata F.	Green semilooper	*Z*5-10:Ac	[18]
Ctenoplusia agnata S.	-	*Z*7-12:Ac	[18]
Zonoplusia ochreata W.	-	*Z*7-12:Ac and *Z*5-12:Ac	[18]
Spodoptera pectinicornis H.	Water lettuce moth	*Z*7-12:Ac	[18]
Gracillariidae			
Phyllocnistis citrella S.	Citrus leaf miner	*Z*7*Z*11*E*13-16:Ald	[19]
Conopomorpha cramerella S.	Cocoa pod borer	*E*4*E*9*Z*10-16:Ac and *E*4*Z*6*Z*10-16:Ac	[20]
Sphingidae			
Agrius convolvuli L.	Convolvulus hawk moth	*E*11*E*13-16:Ald	[21]
Tortricidae			
Homona coffearia N.	Tea tortrix	12:OH	[22]
Archips atrolucens D.	Citrus leaf roller	*Z*11-14:Ac and *E*11-14:Ac	[18, 23]
Adoxophyes privatana W.	Apple leaf-curling moth	*Z*11-14:Ac	[23]
Meridemis furtiva D.	-	*Z*11-14:Ac	[18]
Pyralidae			
Etiella zinckenella T.	Pea pob borer	*Z*11-14:Ac	[24]
Yponomeutidae			
Prays endocarpa M.	Citrus pock caterpillar	*Z*7-14:OH and *Z*7-14:Ald	[25]
Limacodidae			
Parasa lepida C.	Nettle caterpillar	*Z*7,9-10:OH	[26]
Coleoptera			
Brentidae			
Cylas formicarius elegantulus S.	Sweet potato weevil	*Z*3-dodecen-1-yl *E*2-butenoate	[27]

Order, family and species	Common name	Major component of pheromone*	Ref.
Chrysomelidae			
Phyllotreta striolata F.	Striped flea beetle	6*R*,7*S*-Himachala-9,11-diene	[28]
Dryophthoridae			
Cosmopolites sordidus G.	Banana root borer	1*S*3*R*5*R*7*S*-sordidin	[29]
Scarabaeidae			
Oryctes rhinoceros L.	Coconut rhinoceros beetle	Ethyl 4-methyloctanoate	[30]
Curculionidae			
Rhynchophorus ferrugineus O.	Red palm weevil	4*S*,5*S*-Ferrugineol	[31]
Cerambycidae			
Xylotrechus quadripes C.	Coffee white stemborer	2*S*-Hydroxydecan-3-one	[32]
Hemiptera			
Aphididae			
Aphis glycines M.	Soybean aphid	1*R*4a*S*7*S*7a*R*-Nepetalactol	[33]
Pentatomidae			
Nezara viridula L.	Southern green stink bug	*Trans*-1,2-Epoxy-*Z*-α-bisabolene	[34]
Aphididae			
Myzus persicae Sulzer	Green peach aphid	*E*-β-Farnesene	[35]
Brevicoryne brassicae D.	Cabbage aphid	4a*S*7*S*7a*R*-Nepetalactone	[36]
Pseudococcidae			
Planococcus citri R.	Citrus mealybug	Planococcyl acetate	[37]
Pseudococcus comstocki K.	Comstock mealybug	2,6-Dimethyl-1,5-heptadien-3-yl acetate	[38]
Heteroptera			
Dysdercus cingulatus F.	Red cotton bug	*S*-Linalool	[39]
Thysanoptera			
Thripidae			
Thrips palmi K.	Melon thrips	*R*-Lavandulyl 3-methyl-3-butenoate	[40]
Diptera			
Cecidomyiidae			
Orseolia oryzae W.	Asian rice gall midge	2*S*,6*S*-Diaxetoxyheptane	[41]

Z,E: *Z,E*-double bonds; *R,S*: *R,S*-enantiomer carbon; number before hyphen: position of a double bond or enantiomer carbon or epoxy; number after hyphen: carbon number of a straight chain; Ac: acetate, OH: alcohol, and Ald: aldehyde.

Table 1. Overview of popular insect pheromones in Vietnam.

3. Synthesis of pheromones

As discussed above, most of the insect pheromones found in Vietnam possess chiral and olefinic structures. Chiral pheromone is defined as a compound containing at least one asymmetric carbon atom, while olefinic attractant bears one or more double bonds C=C in the carbon chain. Generally, insects are attracted more efficiently by a typical optical or/and configurative isomer than by a mixture of its isomers. Hence, an unambiguous understanding of production of these pheromones is particularly necessary for their application. Herein are summarized important synthetic approaches of pheromones which have been used for field trials in Vietnam over three decades. Synthetic approaches are divided into three main categories as described below.

3.1. Chiral pheromones

3.1.1. 4R,8R-dimethyldecanal (4R,8R-1, 4R,8R-Tribolure, 1)

Suzuki *et al.* [42] described the identification and synthesis of tribolure (**1**) as the aggregation pheromone of the *Tribolium* flour beetles, in which the (4*R*,8*R*)-**1** isomer is a major component of the natural pheromone [43, 44]. A simple way for synthesis of (4*R*,8*R*)-**1** has been reported by Nguyen and co-workers [45–47]. The selective peroxidation of (*S*)-3,7-dimethylocta-1,6-diene (**2**) gave an important intermediate diol **3** which was as an initial material for synthesis of both components, the tosylate **5** and the Grignard reagent **6**. The Wurtz condensation of the two components in presence of lithium cuprate, followed by simple conversions, affords the pheromone (4*R*,8*R*)-**1** (**Scheme 1**).

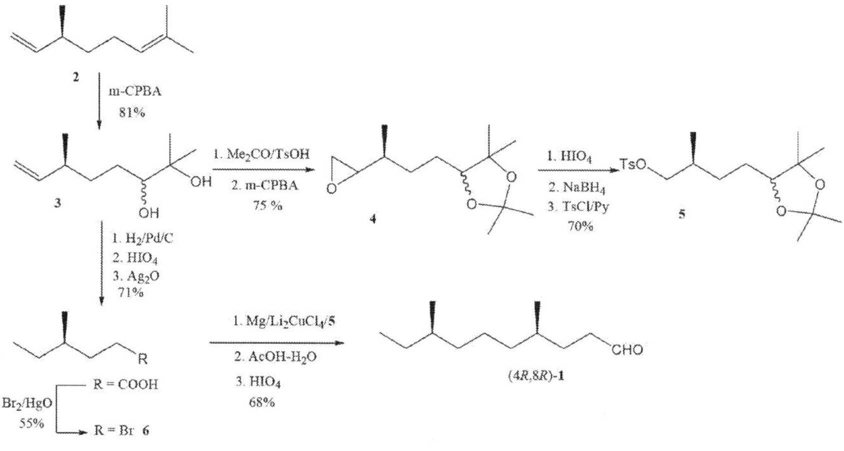

Scheme 1. Synthetic route of 4R,8R-dimethyldecanal.

3.1.2. *(4S, 6S, 7S)-7-hydroxy-4,6-dimethylnonan-3-one (Serricornin, 7)*

(4*S*, 6*S*, 7*S*)-7-hydroxy-4,6-dimethylnonan-3-one (**7**) named Serricornin is the female-produced sex pheromone of the small tobacco beetle, *Lasioderma serricorne,* that has been isolated and identified by Chuman *et al.* [48, 49] The key step in synthesis of this pheromone was reaction between **8** with ethyl triisopropoxytitanium according to Cram's rule to obtain a ratio of isomer (3*S*,4*S*)-**9** with ee 85% [50]. In order to isolate individual optical isomers, (*S*)-1-phenylethylamine (S-PEA) was treated with acid **10** to afford a mixture of diastereomeric salts [51]. After repeated crystallization, the (3*S*, 4*S*)-**10** exhibited ee 92%. The compound **10** was converted into the lactone by a two-step procedure, followed by Grignard coupling that completed synthesis of the target pheromone (**Scheme 2**).

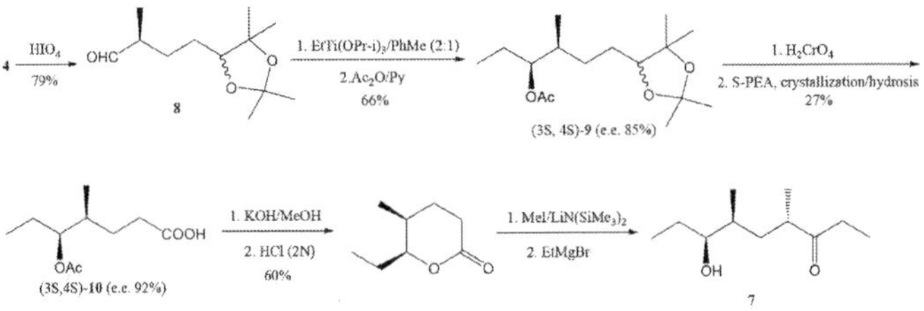

Scheme 2. Synthetic route of Serricornin.

3.1.3. *(10R)-methyldodecyl acetate (10R-11)*

The smaller tea tortrix moth, *Adoxophyes* sp., is a widespread and economically important pest of the tea plant. It has been demonstrated that the male-produced sex pheromone consists of four components, in which 10*R*-**11** was identified as the minor component [52]. Tamaki *et al.* [53] showed that the 10*R*-**11** was more bioactive than the *S*-isomer in field test. The synthesis of *R*-**11** has been reported through two approaches which employed the Grignard coupling

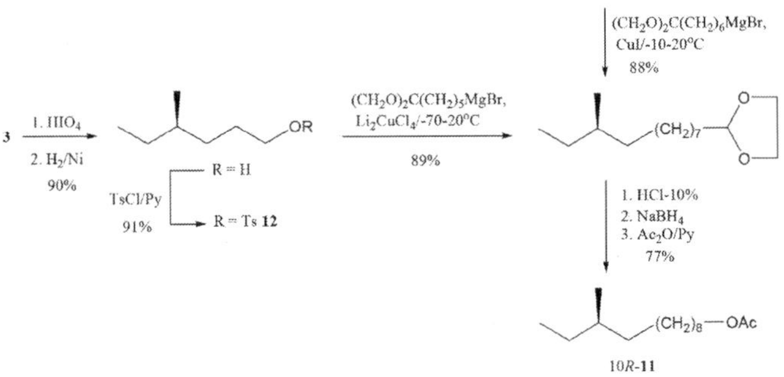

Scheme 3. Synthetic route of (10R)-methyldodecyl acetate.

between the achiral and chiral units as a key step [54]. The latter unit was obtained from the chiral diol **3** converted into the tosylate **12** or the above bromide **6**. Coupling of these compounds with the corresponding Grignard reagents of protected aldehydes and subsequent esterification completed synthesis of the pheromone 10*R*-**11** (**Scheme 3**).

3.2. Racemic pheromones

3.2.1. *Ethyl 4-methyloctanoate (13)*

Hallett *et al.* [30] described identification of the aggregation pheromone of Rhinoceros beetles, *Oryctes rhinoceros* L., the most important destructive pest of coconut, oil, and other palms in tropical Southern Asia, Pacific islands, and Indian islands as ethyl 4-methyloctanoate (**13**). The racemic pheromone [55, 56] has been straightforwardly synthesized from natural citronellol with one-step oxidation of 2,6-dimethyl-2-decene by $KMnO_4$-$FeCl_3$ as the key step, followed by esterification in overall yield of 45% (**Scheme 4**).

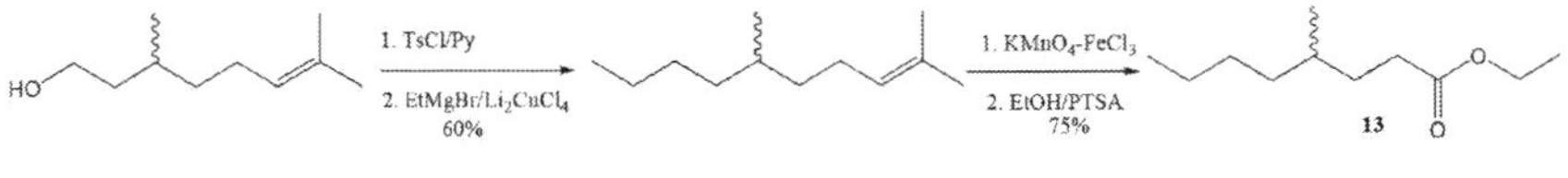

HO … 1. TsCl/Py; 2. EtMgBr/Li_2CuCl_4 60% … 1. $KMnO_4$-$FeCl_3$; 2. EtOH/PTSA 75% … **13**

Scheme 4. Synthetic route of ethyl 4-methyloctanoate.

3.2.2. *5,9-Dimethylpentadecane (14) and 5,9-dimethylhexadecane (15)*

Francke *et al.* [57] identified and synthesized the major and minor components of sex pheromone of female leaf miner moths, *Leucoptera coffeella*, a pest of coffee trees as 5,9-dimethylpentadecane (**14**) and 5,9-dimethylhexadecane (**15**), respectively. Synthesis of these racemic components has been described from citronellol by Doan *et al.* [58]. Grignard coupling reactions with tosylated intermediates under ultrasound irradiation, which has been efficiently employed in literature [59], were the key steps in the synthetic strategy. The alkene derivatives **16** were oxidized and then reduced to afford the important tosylated synthon **17**. Grignard reaction of the corresponding tosylates **17** with 2-methylhexylmagiesium bromide furnished the racemic pheromones, **14** and **15**, under accelerating ultrasound irradiation in yields over 90% (**Scheme 5**).

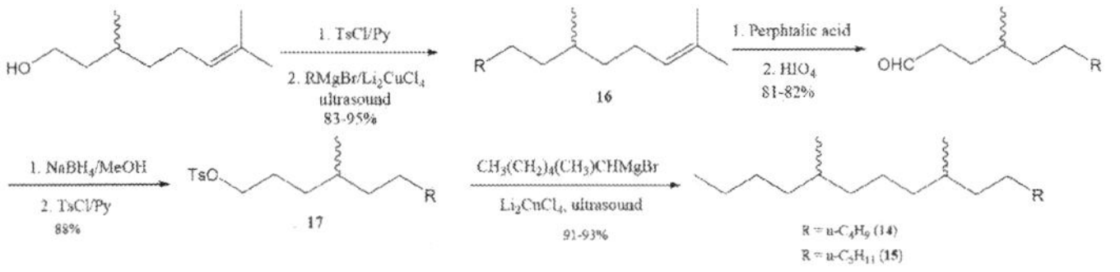

HO … 1. TsCl/Py; 2. RMgBr/Li_2CuCl_4 ultrasound 83-95% … R … **16** … 1. Perphtalic acid; 2. HIO_4 81-82% … OHC … R
1. $NaBH_4$/MeOH; 2. TsCl/Py 88% … TsO … R **17** … $CH_3(CH_2)_4(CH_3)CHMgBr$; Li_2CuCl_4, ultrasound 91-93% … R
R = n-C_4H_9 (14)
R = n-C_5H_{11} (15)

Scheme 5. Synthetic route of 5,9-dimethylpentadecane and 5,9-dimethylhexadecane.

3.2.3. *8-Methyldec-2-yl propanoate (18) and 10-methyl-2-tridecanone (19)*

8-Methyldec-2-yl propanoate (**18**) was identified as sex pheromone of northern corn rootworm, *Diabrotica longicornis* Say [60] and western corn rootworm, *Diabrotica virgifera virgifera* Le Conte [61] while the sex pheromone of southern corn rootworm, *Diabrotica undecimpunctata howardi* Barber, was isolated and identified as 10-methyl-2-tridecanone (**19**) by Guss *et al.* [62]. A method for the synthesis of mixture (2*R*/*S*,8*R*)-**18** with a chiral centre at C-8 was performed using three chiral substrates using Grignard coupling as a key step [63]. The important ketone synthon was reduced with $NaBH_4$, followed by esterification to obtain the pheromone in total yield over 50% (**Scheme 6**).

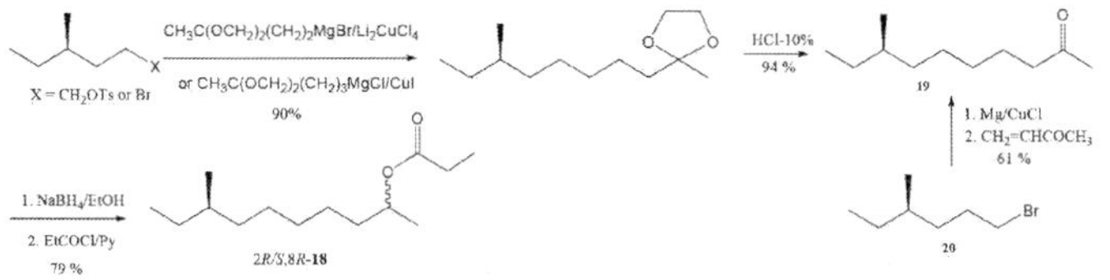

Scheme 6. Synthetic route of 8-methyldec-2-yl propanoate.

In a similar fashion, synthesis of the sex pheromone 10*R*-**19** from the chiral material was based on the successive reaction of 1-tosyloxy-4R-methylheptane and the Grignard reagent of 1-bromo-5,5-ethylenedioxyhexane [64]. A straightforward approach to the synthesis of racemic mixtures of **18** and **19** has been recently reported from diol derivatives using the Grignard coupling of protected bromohydrins as a key step [65]. The important intermediate aldehydes, which have a similar structure in both pheromones, were synthesized by oxidation reaction of corresponding alcohols using PCC as an oxidation reagent. Pheromones **18** and **19** were obtained in overall yields of 35% and 29%, respectively (**Scheme 7**).

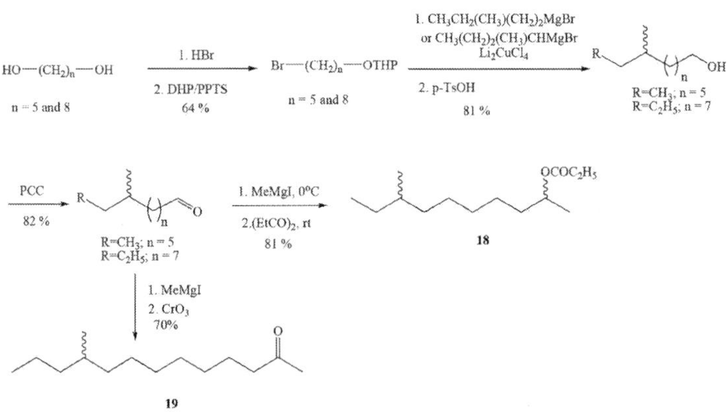

Scheme 7. Synthetic route of 8-methyldec-2-yl propanoate and 10-methyl-2-tridecanone.

3.3. Olefinic pheromones

3.3.1. (14R)-Methyl-8-hexadecenal (14R-21)

The geometrical isomers of (14*R*)-methyl-8-hexadecenal (14*R*-**21**) were identified as major components of the pheromone of Khapra beetle, *Trogoderma granarium*. Both pheromone isomers were found in a *Z*: *E* ratio of 92 : 8 and named (*Z*) and (*E*)-trogodermal [66]. Mori and coworkers [67, 68] demonstrated that the *R*-enantiomers, 14*R*,8*Z*-**21** and 14*R*,8*E*-**21**, revealed bioactivity on male dermestid beetles, *T. glabrum*, *T. inclusum* and *T. variabile*. Nguyen *et al.* [69, 70] described synthesis of the both *R*-isomers based on the chiral bromide substrate **20** and employed substitution coupling as a key step. **Scheme 8** shows synthetic route of the *E*-isomer from acrolein using S_N2' substitution of Grignard reagent of **20** to acyclic allyl acetate to afford the *E*-isomer with purity of 90%, containing a small amount of branched product. At the same time, the Z-isomer was efficiently prepared *via* three steps in overall yield 24% from commercially available *Z*-2-buten-1,4-diol. The key step was condensation of *Z*-disubstituted primary allyl acetate with Grignard reagent according to the nucleophilic substitution mechanism (S_N2) to furnish the pheromone 14*R*,8*Z*-**21** as a major product (**Scheme 9**).

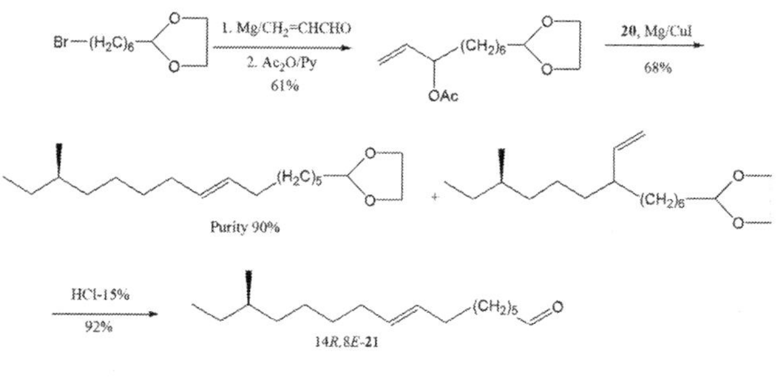

Br—$(H_2C)_6$— 1. Mg/CH_2=CHCHO 2. Ac_2O/Py 61% — $(CH_2)_6$ OAc — **20**, Mg/CuI 68% — $(H_2C)_5$ Purity 90% + $(CH_2)_6$ — HCl-15% 92% — $(CH_2)_5$ O 14*R*,8*E*-**21**

Scheme 8. Synthetic route of (14R, 8E)-14-methyl-8-hexadecenal.

HO OH **22** — 1. $SOCl_2$ 2. $MgCl(CH_2)_5CH(OCH_2)_2$ 3. Ac_2O/Py 34% — AcO $(H_2C)_5$ — **20**, Mg/CuI 71% — $(H_2C)_5$ Purity 93 % + $(CH_2)_6$ — HCl-15% 97% — $(H_2C)_5$ O 14*R*,8*Z*-**21**

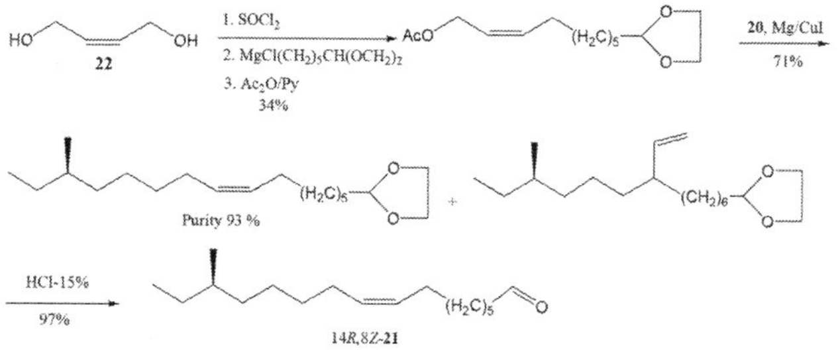

Scheme 9. Synthetic route of (14R, 8Z)-14-methyl-8-hexadecenal.

3.3.2. (Z)-7-Dodecenol (23), (Z)-7-dodecen-1-yl acetate (24), (Z)-7-tetradecenol (25), (Z)-7-tetradecen-1-yl acetate (26), (Z)-7-tetradecenal (27)

Berger *et al.* [71] isolated and identified (Z)-7-dodecen-1-yl acetate (**24**) as the male-produced pheromone of cabbage loopers, *Trichoplusia ni* Hubner, a destructive pest of peas and weed plants in Asia while the precursor of this pheromone, (Z)-7-dodecen-1-ol (**23**) was reported as an inhibitor of this sex pheromone by Tumlinson *et al.* [72]. Vang *et al.* [25] described the identification of three compounds, (Z)-7-tetradecenol (**25**), (Z)-7-tetradecen-1-yl acetate (**26**), and (Z)-7-tetradecenal (**27**) as the sex pheromone of the citrus pock caterpillar, *Prays endocarpa*. The compound **26** has also been found as sex pheromone of other species such as the citrus flower moth (*P. citri* and *P. nephelomina*) [73, 74] and the olive moth (*P. oleae Bern*) [75, 76]. These pheromones have been synthesized from the available commercial diol **22** *via* the S_N2 mechanism between Grignard reagent prepared from protected bromohydrins with Z-allyl acetate derivatives in presence of CuI catalyst as a key step [77]. The Z-isomers were purified by column chromatography using Silica gel impregnated with $AgNO_3$ as the stationary phase (**Scheme 10**).

22 → (1. $SOCl_2$; 2. RMgBr; 3. Ac_2O; 32-43%) → R–CH=CH–CH₂OAc → (1. THPO-$(CH_2)_5$-MgBr/CuI; 2. p-TsOH; 58-60%) → R–CH=CH–$(CH_2)_5$–OH + HO–$(H_2C)_5$–CH(CH=CH₂)–R

R = C_3H_7 (**23**)
R = C_5H_{11} (**25**)

→ (PCC or Ac_2O/Py; 86-90%) → R–CH=CH–$(CH_2)_4$–X

R = C_3H_7, X = CH_2OAc (**24**)
R = C_5H_{11}, X = CH_2OAc (**26**)
R = C_5H_{11}, X = CHO (**27**)

Scheme 10. Synthetic route of derivatives of (Z)-7-dodecenol and (Z)-7-tetradecenol.

3.3.3. (Z)-3-Dodecen-1-yl (E)-2-butenoate (28), (Z)-11-hexadecenol (29), (Z)-11-hexadecenyl acetate (30), (Z)-11-hexadecenal (31)

The sweet potato weevil, *Cylas formicarius elegantulus* S., and Diamondback moth, *Plutella xylostella* L., are prevalently serious insects in Vietnam. Heath *et al.* [27] identified and first synthesized (Z)-3-dodecen-1-yl (*E*)-2-butenoate (**28**) as the female-produced sex pheromone of the sweet potato weevil. The female of Diamondback moth secretes the pheromone to attract the males identified as (Z)-11-hexadecenyl acetate (**30**) and (Z)-11-hexadecenal (**31**) in a ratio of 1:1 to 3:1 [78]. Yamada and Koshihara [79] found (Z)-11-hexadecen-1-ol (**29**) synergizing the attractiveness of the pheromone mixtures of **30** and **31**.

The efficient synthetic pathway of these pheromones from commercially unsaturated acid derivatives has been described [80]. Acrylic acid or 1-undecenic acid was straightforwardly converted into the corresponding triphenylphosphonium salt of methyl esters which reacted with 1-alkanal in presence of dibenzo-18-crown-6 to afford (Z)-ester derivatives. Reduction of these derivatives with $LiAlH_4$ was done to obtain the corresponding alcohols which were

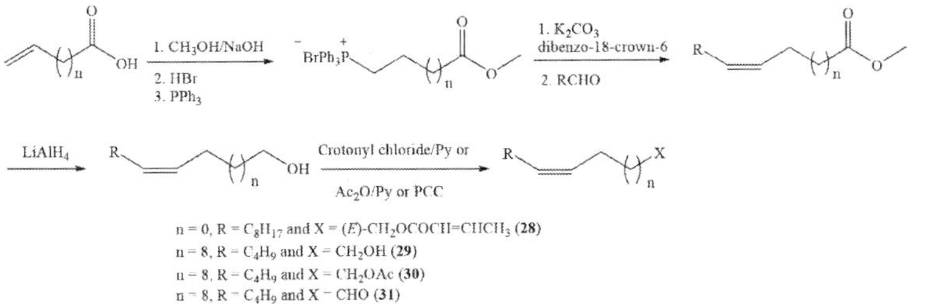

$n = 0$, $R = C_8H_{17}$ and $X = (E)\text{-}CH_2OCOCH{=}CHCH_3$ (**28**)
$n = 8$, $R = C_4H_9$ and $X = CH_2OH$ (**29**)
$n = 8$, $R = C_4H_9$ and $X = CH_2OAc$ (**30**)
$n = 8$, $R = C_4H_9$ and $X = CHO$ (**31**)

Scheme 11. Synthetic route of (Z)-3-dodecen-1-yl (E)-2-butenoate and derivatives of (Z)-11-hexadecenol.

subsequently oxidized with PCC into the aldehyde pheromone (**31**) or esterified with crotonyl chloride and anhydride acetic into pheromones **28** and **30**, respectively (**Scheme 11**). Another way for synthesis of **29–31** based on (Z)-2-buten-1,4-diol similar to the description in **Scheme 10** has also been reported by Nguyen [81].

3.3.4. 7,11,13-Hexadecatrien-1-ol (32) and 7,11,13-hexadecatrienal (33)

Ando *et al.* [82] have identified (*Z*7,*Z*11)-hexadecatrienal (*Z*7,*Z*11-**34**) as an sex attractant of the citrus leaf miner females, *Phyllocnistis citrella* Stainton, a harmful citrus pest in Asia. However, two other research groups [83, 84] demonstrated that a mixture of (*Z*7,*Z*11,*E*13)-hexadecatrienal (*Z*7,*Z*11,*E*13-**33**) and *Z*7,*Z*11-**34** at a ratio of 3:1 strongly attracted the citrus leaf miner in Brazil and California. Vang *et al.* [19] have described synthesis and comparison of biological test of two geometrical isomers, *Z*7,*Z*11,*E*13-**33** and *Z*7,*E*11,*E*13-**34**. The isomers were synthesized, the key steps being Wittig reaction of the protected ylides using a base $NaN(SiMe_3)_2$ to furnish (Z)-isomer as the major products in overall yield of 3%. The pure individual isomers, (*Z*7,*Z*11,*E*13)-**33** and (*Z*7,*E*11,*E*13)-**33,** were obtained by isolation from a mixture of corresponding alcohols **32** (2:1) using preparative HPLC methodology, followed by oxidation with PCC (**Scheme 12**).

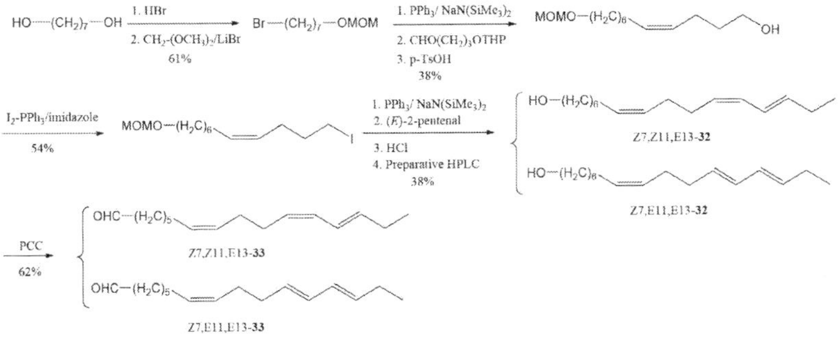

Scheme 12. Synthetic route of isomers of 7,11,13-hexadecatrien-1-ol and 7,11,13-hexadecatrienal.

4. Pheromone trap

Pheromone trap is a useful tool for management of insects. Numerous trap types are being used efficiently for IPM such as board trap, tube trap, and pitfall trap (**Figure 1**). The board traps are most commonly used for trapping moths damaging vegetables or forest trees, while the tube traps are used efficiently for trapping fruit fly. These traps are either hung from branches or nailed to infested trees. Pitfall trap is usually employed to collect beetles and weevils. This trap consists of a container with a lot of small windows buried in the ground where its windows are at surface level [85]. The use of trap generally baited with a lure containing the corresponding pheromone is significantly dependent on individual insect characteristics. The pheromone is dissolved in a suitable solvent, usually hexane and then placed on a rubber septum or septa to protect the compounds from degradation. After evaporation of the solvent, the lure is pinned at centre of the traps which would be put on the field at suitable intervals. Lures are designed to release pheromones at a constant or near-constant rate during the course of the experiment. Release rate is dependent on the molecular, physical and chemical properties of the lure matrix and environmental conditions such as temperature and local weather. Traps are checked and insects are recorded over time, usually for several days or weeks.

Board trap Tube trap Pitfall trap

Figure 1. Examples of trap types.

5. Control of harmful insects in Vietnam

5.1. Diamondback moth, *Plutella xylostella*

- Lure type: synthesized pheromone, **29** (>98%, GC), **30** (99.2%, GC) and **31** (95.5%, GC) in a ratio of 1:10:1 and lures provided from AgriSense Co.
- Trap type: sticky board traps (**Figure 2**).
- Areas: vegetable fields in Tu Liem (HaNoi, north of Vietnam), Da Lat (Lam Dong province, central highland region of Vietnam), Cu Chi and Hoc Mon District (Ho Chi Minh City, South of Vietnam).

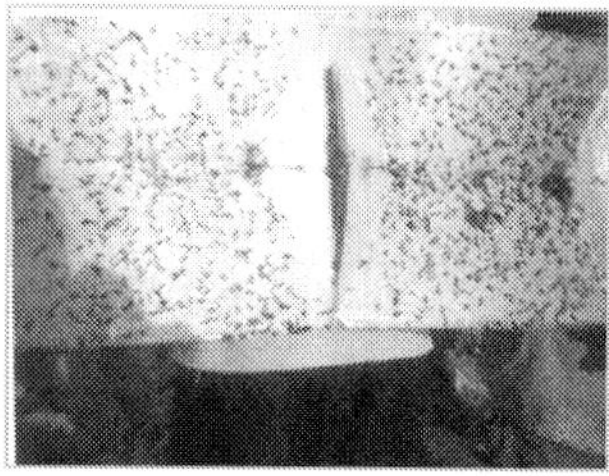

Figure 2. Sticky board trap and captured Diamondback moths, *P. xylostella* [86].

Diamondback moth attacks only cruciferous vegetable crops in the family Cruciferae including cabbage, broccoli, cauliflower, collard, etc. Plant damage is caused by larvae feeding on foliar tissue and this is particularly damaging to the seedlings, resulting in disrupted head formation. A comparison study of field test using the synthesized pheromones and commercial lures was carried out between years 1994 and 1995 in many areas of Vietnam [80]. The summary of the results is presented in **Table 2**. It reveals that distributing density of Diamondback moths is clearly different among geographic areas in Vietnam. For example, the mean males captured by all lures in Da Lat and Cu Chi are significantly higher than the pests in Tu Liem and Hoc Mon district. The authors also demonstrated stronger male response when dosage of the pheromone mixture steadily increases and the captured males by lures containing 5 mg of the synthesized pheromone are similar to the commercial lures.

Lure	Captured mean males/trap/night±S.E.					
(mg)	Tu Liem (Jan. 2 to Feb. 2, 1994)	Da Lat (Apr. 20 to May 20, 1994)	Cu Chi (Mar. 1 to Apr. 1, 1994)	Hoc Mon (Jan. 4 to Feb. 4, 1994)	Cu Chi (Mar. 1 to Apr. 1, 1995)	Hoc Mon (Jan. 4 to Feb. 4,1995)
1[a]	51.9±11.3	95.9±19.9	68.3±15.2	–	–	–
3[a]	95.5±19.2	183.9±25.6	112.2±21.3	–	–	–
5[a]	116.3±21.3	258.3±30.8	204.9±32.5	126.5±16.2	212.4±32.1	116.4±17.1
Com.[b]	117.9±22.6	286.4±35.5	214.6±35.1	131.2±19.7	216.1±33.7	124.8±18.3
Control	1.5	2.1	1.2	–	–	–

[a] Loaded with synthesised pheromone.
[b] Lure of AgriSense Co.

Table 2. Pheromone trap catches of *P. xylostella* males testing different dosages in different areas for one month in years 1994 and 1995 [80].

5.2. Sweet potato weevils, *Cylas formicarius elegantulus*

- Lure type: lures baited with the synthesized pheromone **28** (96%, GC) and products provided by Department of Entomology, CIP Aptartodo 5969, Lima 1, Peru.
- Trap type: tube trap made from plastic flask with six windows (**Figure 3**).

- Area: sweet potato fields in Hoc Mon district (Ho Chi Minh City).

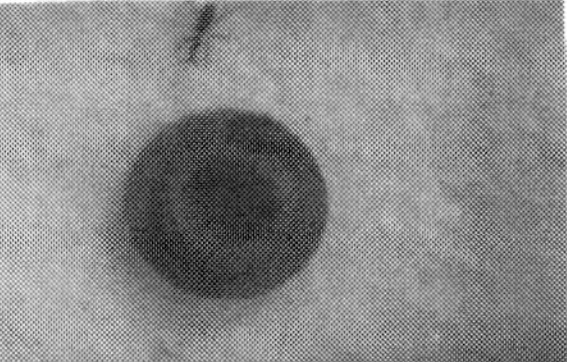

Figure 3. Tube traps and captured sweet potato weevils, *C. formicarius elegantulus* [86].

Sweet potato weevil, *C. formicarius* is the most serious pest of sweet potato which mainly causes damage to fields not only in Vietnam but also around the world. The adult female deposits a single egg near the juncture of stem and tuber and the hatched larvae burrow directly into the tuber of plant. Management of this pest is particularly important in agriculture and storage. Nguyen *et al.* [80] carried out field tests of the weevils in sweet potato fields of Hoc Mon district of Ho Chi Minh City. The results revealed that the weevil capture increased with an increase of pheromone dosage and two pheromone sources, synthesized and commercial, are comparable in their attractiveness to this weevil (**Table 3**).

Cultivation season	Captured mean males/trap/night±S.E				
	Lure (mg)				
	0.01[a]	0.1[a]	1.0[a]	1.0[b]	Control
1	124.6±30.5	215.3±86.4	486.8±183.4	541.2±202.3	0
2	108.9±26.2	202.6±39.1	455.9±121.5	501.6±136.2	0

[a] Loaded with synthesised pheromone.
[b] Pheromone provided from Department of Entomology, CIP Apartado 5969, Lima 1, Peru.

Table 3. Trap catches of *C. formicarius* males testing different dosages of the pheromone for one week in 1995 [86].

5.3. Screening sex attractants for moths in Mekong Delta of Vietnam

- Lure type: synthesized and commercial compounds
  - Monoenyl alcohols: Z7-12:OH (**23**), Z11-14:OH and E11-14:OH
  - Monoenyl acetates: Z5-10:Ac, Z5-12:Ac, Z7-12:Ac (**24**), Z9-12:Ac, Z9-14:Ac, Z11-14:Ac, E11-14:Ac
  - Dienyl acetates: Z9,E11-14:Ac and Z9,E12-14:Ac
  - Monoenyl aldehyde: Z11-14:Ald
  - Trienes: Z3,Z6,Z9-18:H, Z3,Z6,Z9-19:H, Z3,Z6,Z9-20:H and Z3,Z6,Z9-21:H

   - Epoxydienes: racemic mixtures of epoxy3,Z6,Z9-18:H, Z3,epoxy6,Z9-18:H, Z3,Z6,epoxy9-18:H and their C_{19}-C_{21} homologs

- Trap type: sticky board trap (30 × 27 cm)
- Area: orchards of Chinese apple, guava and longan in Can Tho City from December 1998 to November 1999 and orchards of plum and guava from January to December in 2000

Hai *et al.* [18] screened six typical attractants to obtain evaluation of the harmful pest population in Mekong Delta of Vietnam. The field tests found male attraction of nineteen Lepidopteran species including nine taxonomically identified species and ten other taxonomically unidentified species. Compound Z11-14:Ac and its mixture were identified as important attractants of Tortricid species such as *A. privatana, A. atrolucens, M. furtive,* while numerous Noctuid species such as *A. signata, C. eriosoma, C. agnate, C. albostriata, A. ochreata, and S. pectinicornis* were attracted by lures baited with Z7-12:Ac (**24**) as a major component. In addition, the seasonal effect clue for male catch of Noctuid species was also observed. For example, the flights were primarily captured in the dry season, from January to March, and in the latter half of the rainy season, September to December whereas the male catch was rarely observed from April to July. These results could help to depict effective ecological behaviour of the pests in these areas.

5.4. Rhinoceros beetle, *Oryctes rhinoceros*

- Lure Type: synthesized compound **13** and the host material, kairomone, which was extracted from fresh coconut tissue in solvents mixture (ethanol (68%), ethyl acetate (27%) and pentane (5%)).
- Trap type: pitfall traps made from 20-liter plastic buckets with window size about 3 × 8 cm (**Figure 4**).
- Area: coconut fields with the 3–10 years old trees in Hau Giang and Ben Tre province (Mekong Delta).

Figure 4. Pitfall trap and captured rhinoceros beetles, *O. rhinoceros* [77].

The coconut rhinoceros beetle causes extensive damage to economically important wild and plantation palms in Vietnam and Southeast Asia region. The adults eat the leaves and burrow into the crown leading to stunted plant development. Dang *et al.* [55] investigated the beetles in two provinces of Mekong Delta between two seasons, dry and rainy, in years 2004 and 2005 using lures baited with pheromone and kairomone. The authors showed that synergism of kairomone leads to an increased beetle response to their pheromone in all trials and the beetle catches were not too different between the two provinces at the same time. The result revealed that the insect density in the rainy season was higher than in the dry season in both areas (**Table 4**). In addition, steady increase of the beetle catches with increasing release rate of the pheromone was also observed (**Table 5**).

Lure (50 mg)	**Captured mean pests/trap/2 weeks±S.E.**			
	Hau Giang		**Ben Tre**	
	Dry season[a]	**Rainy season**[b]	**Dry season**[c]	**Rainy Seasond**[d]
Kairomone (K)	0	0	0	0
Pheromone (P)	1.33±0.33	1.67±0.33	1.67±0.33	1.67±0.33
P+K	2.67±0.27	6.67±0.27	3.33±0.43	8.67±0.27

[a] Jan. 25, 2014 to Feb. 27, 2004.
[b] June 5, 2014 to June 19, 2004.
[c] Feb. 4, 2005 to Feb. 19, 2005.
[d] June 15, 2004 to June 29, 2004.

Table 4. Trap catches of rhinoceros beetles testing in different areas in dry and rainy reasons of years, 2004 and 2005 [55].

Lure (mg)	30	40	50	60	Control
Pests[a]	0.33±0.33	0.33±0.33	2.33±0.88	3.00±0.58	0

[a]Mean number/trap/2 weeks±S.E.

Table 5. Trap catches of *O. rhinoceros* pests testing different dosages of the pheromone combining with kairomone in Ben Tre province from Apr. 1 to Apr. 15, 2005 [55].

5.5. Citrus leaf miners, *Phyllocnistis citrella*

- Pheromone: two geometrical isomers of synthesized compounds **32** and **33** and commercial isomers of **34**.
- Trap type: sticky board trap (30 × 27 cm).
- Area: citrus orchards in Can Tho City from November 21, 2005 to March 12, 2006 and in Ogasawara Islands, Japan from November 17, 2005 to April 5, 2006.

The citrus leaf miner, a widespread Asian species, is found to be dangerous to all citrus in many areas around the world such as, in East and South Africa. The larvae make serpentine mines on young leaves or shoots, resulting in leaf curling and inducing a serious plant disease. Vang *et al.* [19] described field trials of the moths in citrus orchards in Vietnam and Japan using isomeric mixture of compounds **44–46**. The data in Vietnam revealed that the citrus leaf miner males were not captured by a lure baited only with *Z*7,*Z*11-**46** but were efficiently attracted with a mixture of *Z*7,*Z*11,*E*13-**45** and *Z*7,*Z*11-**46** in a ratio of 3:1, whereas similar synergistic effect on moth catches was not observed in Japan. The authors, therefore, concluded that the sex pheromone of Vietnamese citrus leaf miners is different from their pheromone in Japan.

5.6. Citrus Pock Caterpillar, *Prays endocarpa*

- Pheromone: synthesized compounds **25**, **26** and **27**.
- Trap type: sticky board trap (30 × 27 cm).
- Area: pomelo orchards of villages in Vinh Long province in 2007 and from April 4 to June 10, 2008.

The citrus pock caterpillar is an economically serious pest of the pomelo (*Citrus grandis* L.) orchard in Vietnam and Southeast Asia. The larvae attack and mine into the peel of the fruit which would either develop into tumours or even drop if the attack is in the early stage of development. Vang *et al.* [25] reported the analysis of gas chromatography-mass (GC-MS) from pheromone gland extract of female moths finding three monoenyl-C_{14} derivatives, **25–27** in a ratio of 10:3:10. Field data showed that the male moth was attracted by only aldehyde **27**, while both other compounds were not attractive as well as no synergism with this aldehyde in the adult catches was found. The authors also demonstrated that achieving efficiency with mass trapping experiment was similar to use of a pesticide in suppression of the pest in pomelo orchard.

5.7. Citrus leaf rollers (*Archips atrolucens*, *Adoxophyes privatana* and *Homona* sp.)

- Pheromone: synthetic and commercial compounds Z11-14:Ac, *E*11-14:Ac, *Z*9-14:Ac, Z9-12:Ac and 14:Ac.
- Trap type: sticky board trap (30 × 27 cm).
- Area: orange orchards in Hau Giang province and Can Tho City in 2011 and 2012.

Recently, the natural pheromones of three citrus leaf rollers, *Archips atrolucens*, *Adoxophyes privatana* and *Homona* sp., which have been known as serious defoliators of citrus trees in Vietnam, were identified using analysis techniques GC-EAD and GC-MS [23]. The results showed that three components Z11-14:Ac, *E*11-14:Ac and 14:Ac in a ratio of 64:32:4 were identified as the sex pheromone of *A. atrolucens* while two other couples Z11-14:Ac and Z9-14:Ac (92:8); Z11-14:Ac and Z9-12:Ac (96:4) as the sex pheromones of *A. privatana* and *Homona* sp., respectively. However, their field test revealed that amount of the minor components in tested pheromones of all three species was slightly greater than those in the natural pheromones. For instance, the best male catches were found for *A. atrolucens*, a blending mixture of Z11-14:Ac, E11-14:Ac and 14:Ac at a ratio of 2:1:1; for *A. privatana*, a blending mixture of Z11-14:Ac and Z9-14:Ac at a ratio of 9:1 and similarly for *Homona* sp., a blending mixture of Z11-14:Ac and Z9-12:Ac at a ratio of 9:1.

6. Conclusions

In this chapter, we review the researches related to synthesis and field evaluation of insect pheromones which have been done by Vietnamese researchers or collaborators since 1980s. Pest control by the use of attractant traps promises to obtain an accurately quantitative estimation of insect population. This important information would help farmers to get more effective protection of the plants in Vietnam. In some cases, using the pheromones could reduce damage to the plants. In addition, pheromones were made possible by advances in synthetic organic chemistry and most of the synthetic approaches mentioned earlier are novel and employable to prepare at a large scale. Some schemes of the synthetic routes have become a standard methodology for synthesis of the similarly structured molecules such as synthesis of enantiomers from (*S*)-3,7-dimethylocta-1,6-diene or synthesis of *Z* and *E*-monoenyl from (*Z*)-2-buten-1,4-diol and acrolein, respectively. Moreover, this study is motivated by the prospect of gaining better understanding of ecological relationships to develop IPM in Vietnam. Also, it helps organic chemists, entomologists and authorities to get sharp orients for further projects in developing an efficient environmentally benign tool. We believe that pheromones are going to become essential materials for the durable development of a green agriculture in the near future.

Author details

Chi-Hien Dang[1], Cong-Hao Nguyen[1], Chan Im[2] and Thanh-Danh Nguyen[1,2*]

*Address all correspondence to: danh5463bd@yahoo.com

1 Department of Technology for Pharmaceutical Chemistry, Institute of Chemical Technology, Vietnam Academy of Science and Technology, HoChiMinh City, Vietnam

2 Department of Chemistry, Konkuk University, Seoul, Republic of Korea

References

[1] Gill HK, Garg H. Pesticide: environmental impacts and management strategies. In: Solenski S, Larramenday ML (eds.). Pesticides—Toxic Effects. Intech. Rijeka, Croatia; 2014. pp. 187–230.

[2] Baker CT. Integrated pest management, use of pheromones in IPM. In: Radcliffe EB, Hutchison WD, Cancelado RE. Integrated Pest Management Concepts, Tactics, Strategies and Case Studies. Cambridge University. New York, USA; 2009. p. 273–285. DOI: 10.1017/CBO9780511626463.022

[3] Nguyen TCH. Insect Pests and Mites of Fruit Plants in the Mekong Delta of Vietnam and Their Management. The Agriculture: HoChiMinh City. 2000. 267 p.

[4] Nguyen CH, Nguyen MT, Pham TT. Some derivatives of propenylbenzene and their attractiveness to fruitfly *Dacus dorsalis* Hendel. Vietnam J. Chem. 1982; 20: 22–26.

[5] Kang SK, Moon BH, Lee JO, Goh HG. A convergent synthesis of (Z)-13-octadecen-1-yl acetate, the pheromone mimic of the rice leaf folder moth and its biological activity test. Bull. Kor. Chem. Soc. 1985; 6: 228–230.

[6] Kawazu K, Hasegawa J, Honda H, Ishikawa Y, Wakamura S, Sugie H, Kamiwada H, Kamimuro T, Yoshiyasu Y, Tatsuki S. Geographical variation in female sex pheromones of the rice leaffolder moth, *Cnaphalocrocis medinalis*: identification of pheromone components in Japan. Entomol. Exp. Appl. 2000; 96: 103–109.

[7] Ohta K, Tatsuk S, Uchiumi K, Kurihara M, Fukam J. Structures of sex pheromones of rice stem borer. Agric. Biol. Chem. 1976; 40: 1897–1899.

[8] Arai K, Ando T, Sakurai A, Yamada H, Koshihara T, Takahashi N. Identification of the female sex pheromone of the cabbage webworm. Agric. Biol. Chem. 1982; 46: 2395–2397.

[9] Usui K, Kurihara M, Uchiumi K, Fukami J, Tatsuki S. Sex pheromone of the female large cabbage-heart caterpillar, *Crocidolomia binotalis* Zeller (Lepidoptera: Pyralidae). Agric. Biol. Chem. 1987; 51: 2191–2195.

[10] Konno Y, Arai K, Sekiguchi K, Matsumoto Y. (E)-10-Hexadecenal, a sex pheromone component of the yellow peach moth, *Dichocrocis punctiferalis* Guenée (Lepidoptera: Pyralidae). Appl. Entomol. Zool. 1982; 17: 207–217.

[11] Permana AD, Samoedi D, Zagatti P, Malosse C. Identification of pheromone components of sugar cane moth top borer, *Tryporiza nivella* F. (Lepidoptera: Pyralidae). Bull. Pusat Pen. 1995; 41: 80–85.

[12] Wu DM, Yan Y., Kong F, Li W, Yu S, Li J, Zeng W. Sex pheromone of the sugarcane striped borer—identification of three components. Youji Huaxue. 1990; 10: 47–49.

[13] Nesbitt BF, Beevor PS, Hall DR, Lester R, Dyck VA. Identification of the female sex pheromone of the purple stem borer moth, *Sesamia inferens*. 1976; 6: 105–107.

[14] Piccardi P, Capizzi A, Cassani G, Spinelli P, Arsura E, Massardo P. A sex pheromone component of the old world bollworm *Heliothis armigera*. J. Insect Physiol. 1977; 23: 1443–1445.

[15] Brady UE, Ganyard MC. Identification of a sex pheromone of the female beet armyworm, *Spodoptera exigua*. Ann. Entomol. Soc. Am. 1972; 65: 898–899.

[16] Tamaki Y, Noguchi H, Yushima T. Sex pheromone of *Spodoptera litura* (F.) (Lepidoptera: Noctuidae): isolation, identification, and synthesis. Appl. Entomol. Zool. 1973; 8: 200–203.

[17] Benn MH, Galbreath RA, Holt VA, Young H, Down G, Priesner E. The sex pheromone of the silver Y moth *Chrysodeixis eriosoma* (Doubleday) in New Zealand. Z. Naturforsch. C. 1982; 37: 1130–1135. DOI: 10.1515/znc-1982-11-1214

[18] Hai TV, Vang LV, Son PK, Inomata SI, Ando T. Sex attractants for moths of Vietnam: Field attraction by synthetic lures baited with known Lepidopteran pheromone. J. Chem. Ecol. 2002; 28: 1473–1481.

[19] Vang LV, Islam MDA, Do ND, Hai TV, Koyano S, Okahana Y, Ohbayashi N, Yamamoto M, Ando T. 7,11,13-Hexadecatrienal identified from male moths of the citrus leafminer as a new sex pheromone component: synthesis and field evaluation in Vietnam and Japan. J. Pestic. Sci. 2008; 33: 152–158 DOI: 10.1584/jpestics.G07-40

[20] Beevor PS, Cork A, Hall DR, Nesbitt BF, Day RK, Mumford JD. Components of female sex pheromone of cocoa pod borer moth, *Conopomorpha cramerella*. J. Chem. Ecol. 1986; 12: 1–23.

[21] Wakamura S, Yasuda T, Watanabe M, Kiguchi K, Shimoda M, Ando T. Sex pheromone of the sweetpotato hornworm, *Agrius convolvuli* (L.) (Lepidoptera: Sphingidae): identification of a major component and its activity in a wind tunnel. Appl. Entomol. Zool. 1996; 31: 171–174.

[22] Kochansky JP, Roelofs WL, Sivapalan P. Sex pheromone of the tea tortrix moth (*Homona coffearia* Neitner). J. Chem. Ecol. 1978; 4: 623–631.

[23] Vang LV, Thuy HN, Khanh CNQ, Son PK, Yan Q, Yamamoto M, Jinbo U, Ando T. Sex pheromones of three citrus leafrollers, *Archips atrolucens*, *Adoxophyes privatana*, and *Homona* sp., inhabiting the Mekong Delta of Vietnam. J. Chem. Ecol. 2013; 39: 783–789.

[24] Tóth M, Löfstedt C, Hansson BS, Szöcs G, Farag AI. Identification of four components from the female sex pheromone of the lima-bean pod borer, *Etiella zinckenella*. Entomol. Exp. Appl. 1989; 51: 107–112.

[25] Vang LV, Do ND, An LK, Son PK, Ando T. Sex pheromone components and control of the Citrus Pock Caterpillar, *Prays endocarpa*, found in the Mekong Delta of Vietnam. J. Chem. Ecol. 2011; 37: 134–140. DOI: 10.1007/s10886-010-9883-2

[26] Wakamura S, Tanaka H, Masumoto Y, Sawada H, Toyohara N. Sex pheromone of the blue-striped nettle grub moth *Parasa lepida* (Cramer) (Lepidoptera: Limacodidae): identification and field attraction. Appl. Entomol. Zool. 2007; 42: 347–352.

[27] Heath RR, Coffelt JA, Sonnet PE, Proshold FI, Dueben B, Tumlinson JH. Identification of sex pheromone produced by female sweetpotato weevil, *Cylas formicarius elegantulus* (Summers). J.Chem. Ecol. 1986; 12: 1489–1503. DOI: 10.1007/BF01012367

[28] Beran F, Mewis I, Srinivasan R, Svoboda J, Vial C, Mosimann H, Boland W, Büttner C, Ulrichs C, Hansson BS, Reinecke A. Male *Phyllotreta striolata* (F.) produce an aggregation pheromone: identification of male-specific compounds and interaction with host plant volatiles. J. Chem. Ecol. 2011; 37: 85–97.

[29] Beauhaire J, Ducrot PH, Malosse C, Rochat D, Ndiege IO, Otieno DO. Identification and synthesis of sordidin, a male pheromone emitted by *Cosmopolites sordidus*. Tetrahedron Lett. 1995; 36: 1043–1046.

[30] Hallett RH, Perez AL, Gries G, Gries R, Pierce JHD, Yue J, Oehlschlager AC, Gonzalez LM, Borden JHL. Aggregation pheromone of coconut Rhinoceros beetle, *Oryctes Rhinoceros* (L.) (Coleoptera: Scarabaeidae). J. Chem. Ecol. 1995; 21: 1549–1570.

[31] Hallett RH, Gries G, Gries R, Borden JH, Czyzewska E, Oehlschlager AC, Pierce HDJ, Angerilli NPD, Rauf A. Aggregation pheromones of two Asian palm weevils, *Rhynchophorus ferrugineus* and *R. vulneratus*. Naturwissenschaften. 1993; 80: 328–331.

[32] Hall DR, Cork A, Phythian SJ, Chittamuru S, Jayarama BK, Venkatesha MG, Sreedharan K, Vinod-Kumar PK, Seetharama HG, Naidu R. Identification of components of male-produced pheromone of coffee white stemborer, *Xylotrechus quadripes*. J. Chem. Ecol. 2006; 32: 195–219.

[33] Zhu JW, Zhang A, Park KC, Baker T, Lang B, Jurenka R, Obrycki JJ, Graves WR, Pickett JA, Smiley D, Chauhan KR, Klun JA. Sex pheromone of the soybean aphid, *Aphis glycines* Matsumura, and its potential use in semiochemical-based control. Environ. Entomol. 2006; 35: 249–257.

[34] Baker R, Borges M, Cooke NG, Herbert RH. Identification and synthesis of (Z)-(1′*S*,3′*R*, 4′*S*)(-)-2-(3′,4′,-epoxy-4′-methylcyclohexyl)-6-methylhepta-2,5-diene, the sex pheromone of the southern green stinkbug, *Nezara viridula* (L.). J. Chem. Soc. Chem. Commun. 1987; 6: 414–416.

[35] Edwards LJ, Siddall JB, Dunam LL, Uden P, Kislow CJ. *Trans*-β-farnesene, alarm pheromone of the green peach aphid, *Myzus perssicae* (Sulzer). Nature. 1973; 241: 126–127.

[36] Gabrys BJ, Gadomski HJ, Klukowski Z, Pickett JA, Sobota GT, Wadhams LJ, Woodcock CM. Sex pheromone of cabbage aphid *Brevicoryne brassicae*: identification and field trapping of male aphids and parasitoids. J. Chem. Ecol. 1997; 23: 1881–1890.

[37] Bierl-Leonhardt BA, Moreno DS, Schwarz M, Fargerlund J, Plimmer JR. Isolation, identification and synthesis of the sex pheromone of the citrus mealybug, *Planococcus citri* (Risso). Tetrahedron Lett. 1981; 22: 389–392.

[38] Negishi T, Uchida M, Tamaki Y, Mori K, Isiwatari T, Asano S, Nakagawa K. Sex pheromone of the comstock mealybug, *Pseudococcus comstocki* Kuwana: isolation and identification. Appl. Entomol. Zool. 1980; 15: 328–333.

[39] Rudmann AA, Aldrich JR. Chirality determinations for a tertiary alcohol: ratios of linalool enantiomers in insects and plants. J. Chromatogr. 1987; 407: 324–329.

[40] Akella SVS, Kirk WDJ, Lu Y, Murai T, Walters KFA, Hamilton JGC. Identification of the aggregation pheromone of the Melon Thrips, *Thrips palmi*. PLoS One. 2014; 9: e103315 DOI:10.1371/journal.pone.0103315

[41] Hall DR, Amarawardana L, Cross JV, Francke W, Boddum T, Hillbur Y. The chemical ecology of *Cecidomyiid midges* (Diptera: Cecidomyiidae). J. Chem. Ecol. 2012; 38: 2–22.

[42] Suzuki T. 4,8-Dimethyldecanal: The aggregation pheromone of the flour beetles, *Tribolium castaneum* and *T. confusum* (Coleoptera: Tenebrionidae). Agric. Biol. Chem. 1980; 44: 2519–2520 DOI:10.1271/bbb1961.44.2519

[43] Lu Y, Beeman RW, Campbell JF, Park Y, Aikins MJ, Mori K. Anatomical localization and stereoisomeric composition of *Tribolium castaneum* aggregation pheromones. Naturwissenschaften. 2011; 98: 755–761. DOI 10.1007/s00114-011-0824-x

[44] Akasaka K, Tamogami S, Beeman RW, Mori K. Pheromone synthesis. Part 245: synthesis and chromatographic analysis of the four stereoisomers of 4,8-dimethyldecanal, the male aggregation pheromone of the red flour beetle, *Tribolium castaneum*. Tetrahedron. 2011; 67: 201–209. DOI:10.1016/j.tet.2010.10.086

[45] Nguyen CH, Mavrov MV, Serebryakov EP. Synthesis of 4R,8R-dimethyldecanal, the aggregation pheromone of flour beetles. Zh. Org. Khim. 1987; 23: 1649–1653.

[46] Nguyen CH, Mavrov MV, Serebryakov EP. Synthesis of 4*R*,8*R*-dimethyldecanal, the aggregation pheromone and analogues. Bioorg. Khim. 1989; 15: 123–126.

[47] Nguyen CH. Synthesis of 4,8-dimethyldecanal from citronellol. Vietnam J. Chem. 1987; 25: 1–3.

[48] Chuman T, Kohno M, Kato K, Noguchi M. 4,6-Dimethyl-7-hydroxynonan-3-one, a sex pheromone of cigarette beetle (*Lasioderma serricorne* F.). Tetrahedron Lett. 1979; 25: 2361–2364. DOI: 10.1016/S0040-4039(01)93974-7

[49] Chuman T, Mochizuki K, Mori M, Kohno M, Kato K, Noguchi M. Lasioderma chemistry sex pheromone of cigarette beetle (*Lasiodema serricorne* F.). J. Chem. Ecol. 1985; 11: 417–434. DOI: 10.1007/BF00989553

[50] Nguyen CH, Mavrov MV, Serebryakov EP. Coleoptera Pheromones. 11. Formal synthesis of Serricornine from S-(+)-3,7-dimethyyl-1,6-octadiene. Rus. Chem. Bull. 1989; 38: 1243–1246 DOI: 10.1007/BF00957161

[51] Serebryakov EP, Nguyen CH, Mavrov MV. Chiral building blocks based on technical grade β-citronellene. Pure Appl. Chem. 1990; 62: 2041–2046.

[52] Tamaki Y, Noguchi H, Sugie H, Sato R, Kariya A. Minor components of the female sex-attractant pheromone of the smaller tea tortrix moth (lepidoptera: tortricidae): isolation and identification. Appl. Ent. Zool. 1979; 14: 101–113.

[53] Tamaki Y, Sugie H, Kariya A, Arai S, Ohba M, Terada T, Suguro T, Mori K. Four-component synthetic sex pheromone of the smaller tea tortrix moth: field evaluation of its potency as an attractant for male moth. Jpn. J. Appl. Entomol. Zool. 1980; 24: 221–228.

[54] Nguyen CH, Mavrov MV, Serebryakov EP. Terpenes in organic synthesis. 6. Synthesis of (*R*)-(−)-10-methyldodecyl acetate from (*S*)-(+)-3,7-dimethylocta-1,6-diene. Rus. Chem. Bull. 1988; 37: 589–591 DOI: 10.1007/BF00965386

[55] Dang CH, Nguyen CH, Nguyen CTHG. The aggregation pheromone of rhinoceros beetles (*Oryctes rhinoceros* L.), synthesis and application in Vietnam. Vietnam J. Sci. Technol. 2008; 46 (4A): 121–129.

[56] Nguyen CH, Dang CH, Nguyen CTHG. Synthesis of ethyl 4-methyloctanoate, the aggreation pheromone of Rhinoceros beetles (*Oryctes rhinoceros* Linn.). Vietnam J. Chem. 2003; 41: 125–130.

[57] Francke W, Toth M, Szocs G, Krieg W, Ernst H, Buschmann E. Identification and synthesis of dimethylalkanes as sex attractants of female leaf miner moths (Lyonetiidae). Z. Naturforsch. 1988; 43c: 787–789.

[58] Doan NN, Le NT, Nguyen CH, Hansen PE, Duus F. Ultrasound assisted synthesis of 5,9-dimethylpentadecane and 5,9-dimethylhexadecane-the sex pheromones of *Leucoptera coffeella*. Molecules. 2007; 12: 2080–2088. DOI:10.3390/12082080

[59] Dang CH, Nguyen CH, Nguyen CTHG, Le TS, Nguyen TD. Applying low ultrasonic in Grignard reactions for synthesis of the insects' pheromone and attractants. In: Proceedings of 12th Asian Chemical Congress. August 23–25, 2007; Kuala Lumpur, Malaysia; 2007. p. 115–123.

[60] Krysan JL, Wilkin PH, Tumlinson JH, Sonnet PE, Carney RL, Guss PL. Responses of *Diabrotica lemniscata* and *D. longicornis* (Coleoptera: Chrysomelidae) to stereoisomers of 8-methyl-2-decyl-propanoate and studies on the pheromone of *D. longicornis*. Ann Entomol Soc Am. 1986; 79: 742–746.

[61] Guss PL, Tumlinson JH, Sonnet PE, Proveaux AT. Identification of a female-produced sex pheromone of the western corn rootworm. J. Chem. Ecol. 1982; 8: 545–556. DOI: 10.1007/BF00987802

[62] Guss PL, Tumlinson JH, Sonnet PE, McLaughlin JR. Identification of a female-produced sex pheromone from the southern corn rootworm, *Diabrotica undecimpunctata howardi* Barber. J. Chem. Ecol. 1983; 9: 1363–1375. DOI: 10.1007/bf00994805

[63] Nguyen CH, Mavrov MV, Serebryakov EP. Coleopteran Pheromone. Communication 9. Synthesis of the attractant of the leaf beetle *Diabrotica virgifera virgifera* (Coleoptera: Chrysamelidae) from (+)-3,7-dimethyl-1,6-octadiene. Rus. Chem. Bull. 1987; 36: 1931–1934. DOI: 10.1007/BF00958349

[64] Nguyen CH, Mavrov MV, Serebryakov EP. Coleopteran Pheromones. Communication 10. Synthesis of the chiral pheromone of the eleven-spotted leaf beetle (Coleoptera: Chrysamelidae). Rus. Chem. Bull. 1987; 36: 1934–1937. DOI: 10.1007/BF00958350

[65] Nguyen TD, Nguyen CH, Im C, Dang CH. Synthesis of corn rootworm pheromones from commercial diols. Chem Pap. 2015; 69: 380–384. DOI: 10.1515/chempap-2015-0027

[66] Levinson AR, Levinson HZ, Schwaiger H, Cassidy RF, Silverstein RM. Olfactory behavior and receptor potentials of the Khapra beetle *Trogoderma granarium* (Coleoptera: Dermestidae) induced by the major components of its sex pheromone, certain analogues, and fatty acid esters. J. Chem. Ecol. 1978; 4: 95–108. DOI: 10.1007/BF00988264

[67] Mori K, Levinson HZ. The pheromone activity of chiral isomers of Trogodermal for male Khapra beetles. Naturwissenschaften. 1980; 67: 148–149. DOI: 10.1007/BF01073624

[68] Silverstein RM, Cassidy RF, Burkholder WE, Shapas TJ, Levinson HZ, Levinson AR, Mori K. Perception by Trogoderma species of chirality and methyl branching at a site far removed from a functional group in a pheromone component. J. Chem. Ecol. 1980; 6: 911–917. DOI: 10.1007/BF00990475

[69] Nguyen CH, Mavrov MV, Serebryakov EP. Pheromones of Coleoptera. Communication 5. Synthesis of *R*, *E*-14-methyl-8-hexadecenal (*trans*-Trogodermal). Rus. Chem. Bull. 1987; 36: 830–833. DOI: 10.1007/BF00962332

[70] Nguyen CH, Mavrov MV, Serebryakov EP. Pheromones of Coleoptera. Communication 6. Synthesis of R,Z-14-methyl-8-hexadecenal (cis-Trogodemal). Rus. Chem. Bull. 1987; 36: 833–837. DOI: 10.1007/BF00962333

[71] Berger RS. Isolation, identification, and synthesis of the sex attractant of the cabbage looper, *Trichoplusia ni*. Ann Entomol Soc Am. 1966; 59: 767–771. DOI: 10.1093/aesa/59.4.767

[72] Tumlinson JH, Mitchell ER, Browner SM, Mayer MS, Green N, Hines R, Lindquist DA. *Cis*-7-dodecen-1-ol, a potent inhibitor of the cabbage looper sex pheromone. Environ. Entomol. 1972; 1: 354–358. DOI:10.1093/ee/1.3.354

[73] Nesbitt BF, Beevor PS, Hall DR, Lester R, Sternlicht M, Goldenberg S. Identification and synthesis of the female sex pheromone of the citrus flower moth, *Prays citri*. Insect Biochem. 1977; 7: 355–359.

[74] Gibb AR, Jamieson LE, Suckling DM, Ramankutty P, Stevens PS. Sex pheromone of the citrus flower moth *Prays nephelomina*: pheromone of the citrus flower moth Prays nephelomina: pheromone identification, field trapping trials, and phenology. J. Chem. Ecol. 2005; 31: 1633–1644.

[75] Campion DG, McVeigh LJ, Polyrakis J, Michaelakis S, Stravrakis GN, Beevor PS, Hall DR, Nesbitt BF. Laboratory and field studies of the female sex pheromone of the olive moth, *Prays oleae*. Experientia. 1979; 35: 1146–1147.

[76] Renou M, Descoins C, Priesner E, Gallois M, Lettere M. Z-7-Tetradecenal, the main component of the sex pheromone of the olive moth *Prays oleae*. C. R. Hebd. Seances Acad. Sci. Ser. D. 1979; 288: 1559–1592.

[77] Dang CH. Synthesis and application pheromone of some harmful insects on fruit-trees in Vietnam [thesis]. Vietnam Academy of Science and Technology, Hanoi. 2008.

[78] Chow YS, Chiu SC, Chien CC. Demonstration of sex pheromone of the diamondback moth (Lepidoptera: Plutellidae). Ann. Entomol. Soc. Am. 1974; 67: 510–512.

[79] Yamada H, Koshihara T. Attractant activity of female sex pheromone of diamondback moth, *Plutella xyloslella* (L.) and analogue. Jpn. J. Appl. Entomol. Zool. 1980; 24: 6–12.

[80] Nguyen CH, Nguyen CTHG, Nguyen CK, Nguyen TS. Synthesis and application of insect attractants in Vietnam. Resources, Conservation and Recycling. 1996; 18: 59–68. DOI: 10.1016/S0921-3449(96)01168-8

[81] Nguyen CH. Synthesis of (Z)-alken and derivatives from 2-butyn-1,4-diol. Vietnam J. Chem. 1985; 23: 8–13.

[82] Ando T, Taguchi K, Uchiyama M, Ujiye T, Kuroko H. (7Z, 11Z)-7,11-Hexadecadienal: sex attractant of the citrus miner moth, *Phyllocnistis citrella Stainton* (Lepidoptera, Phyllocnistidae). Agric. Biol. Chem. 1985; 49: 1633–3635.

[83] Leal WS, Parra-Pedrazzoli AL, Cosse AA, Murata Y, Bento JMS, Vilela EF. Identification, synthesis, and field evaluation of the sex pheromone from the citrus leafminer, *Phyllocnistis citrella*. J. Chem. Ecol. 2006; 32: 155–168. DOI: 10.1007/s10886-006-9358-7

[84] Moreira JA, Mcelfresh JS, Millar JG. Identification, synthesis, and field testing of the sex pheromone of the citrus leafminer, *Phyllocnistis citrella*. 2006; 32: 169–194. DOI: 10.1007/s10886-006-9359-6

[85] Gill HK, McSorley R, Buss L. 2010. The insect community on the soil surface (http://edis.ifas.ufl.edu/in876) date of access: 15/4/2016.

[86] Nguyen CH. Studied on Application of Insect Hormones in the Agriculture. The Agriculture: HoChiMinh City. 2009. 316p.

Chapter 6

Using Semiochemicals for Coleopterean Pests in Sustainable Plant Protection

Maria Pojar-Fenesan and Ana Balea

Additional information is available at the end of the chapter

http://dx.doi.org/10.5772/64278

Abstract

Since the early twentieth century, chemical industry provides farmers large amounts of synthetic chemicals used as fertilizers and pest control products. Agriculture became intensive and the crop yields and the profit increased dramatically. Tons of toxic material for crop protection and fertilization were scattered through the gardens, fields, and orchards. But all these chemicals affect the environment, with serious negative consequences for humanity, today and tomorrow. Since 1959 until today many researchers observed and discussed the disadvantages of chemical methods of combating harmful insects and misapplied their disruptive action on cultivated ecosystems. Integrated pest management (IPM) is a process used to solve pest problems with minimum risks to people and environment. The objective of the researches presented in this chapter is to obtain and testing "*semio*chemicals"—pheromones, involved in intraspecific communication, into environmentally compatible strategies, to reduce pest populations under economic damage thresholds. Insect that are monitored and mass trapped in proposed IPM strategies are potato pest, Colorado potato beetle, *Lepidoptera decemlineata* Say (Coleoptera, Chrisomelidae); corn pests, West Corn Rootworm *Diabrotica virgifera virgifera* (Coleoptera, Chrisomelidae); and six-spined bark beetle *Pityogenes chalcographus* (Coleoptera, Scolitydae) pest in coniferous forests.

Keywords: semiochemicals, sexual, aggregation pheromones, kairomones, control, pest, Coleoptera, Colorado potato beetle (*Leptinotarsa decemlineata* SAY), western corn rootworm (*Diabrotica virgifera virgifera*), bark beetle, *Pityogenes chalcographus*

1. Introduction

The care for the next generations involves the resources protection, keeping soil, air, and waters clean. In agriculture, for a sustainable future, scientists and farmers must develop the environmentally friendly, economically viable, and socially responsible technologies [1]. Intensive agriculture uploads environment with pollutants such as pesticides, herbicides, and fertilizer. In the European Union, 38% of bird species and 45% of lepidopteran species are threatened with extinction. It is true that the high populations of insects' pests could destroy the human food sources, and therefore need to be maintained under economic damage level, but each organism has its role in the ecosystem to which it belongs. The pesticides kill not only harmful insects but also beneficial organisms, and thus the ecosystem equilibrium is modified. The most affected are the pollinators such as honeybees.

The concept of integrated protection or integrated pest management (IPM) appeared at the end of the sixth decade of twentieth century by the works of Dutch researcher, Briejer [2] and Americans: Smith and Hagen [3] and Stern and van den Bosch [4]. The disadvantages of chemical methods of combating harmful insects and misapplied their disruptive action on cultivated ecosystems were discussed recently by Gill and Garg [5].

Modern ecofriendly crop protection strategies are discussed in symposiums organized by Food and Agriculture Organization of the United Nations (FAO) and IOBC/WPRS (International Organization for Biological and Integrated Control of Noxious Animals and Plants/West Regional Paleartic Section). According to the FAO, IPM means considering all available pest control techniques and other measures that reduce the development of pest populations, with minimum risks to human health and the environment.

Development of pest management alternatives based on mediators' chemicals has been necessitated by the loss of traditional pesticides, insect pest resistance, pest resurgences, and secondary pest outbreaks often due to the effect of pesticides on all environments [5].

Semiochemicals, defined as behavior-modifying chemicals, are volatile organic compounds that transmit chemical messages, "words" in organism "language" and are used by insects for intra- and interspecies communication. The term "*semio*chemical" derived from the Greek word "semeon," which means "sign" or "signal" [6]. Insects detect volatiles semiochemicals directly from the air with olfactory receptors located in sensilla hairs on the antennae and the effect is a change in insect behavior. Semiochemicals can be classified as pheromones or allelochemicals based on how they are used and who benefits [7].

The intraspecific communication language have as "words" volatile signals, so called pheromones, emitted by an organism that produces on the receptors of the same species a behavioral change. The term "pheromone" is derived from the Greek words "pherein" (to carry) and "hormone" (to stimulate), and was introduced by Karlson and Butenandt [8]. Based on their effect, pheromones categories are as follows:

- Aggregation pheromones: signaling an important place for the life, e.g., where insects' species could find the "food" or could lay eggs.

- Alarm pheromones: compounds that stimulate insects' escape or defense behavior.
- Sex pheromones: emitted by the female (in most of the cases) inducing male of the same species mating behavior.
- Trail pheromones: social insects as workers ants released pheromones to mark the way to a food source.
- Marking pheromones: compounds used by insects to mark the territory.

The allelochemicals are classified as allomones, kairomones, or synomones [7]. Allomones are a class of compounds that benefit the producer, but not the receiver. Allomones are often used for defense, such as toxic insect secretions. Predators also use allomones to lure prey. Kairomones are a class of compounds that are advantageous for the receiver. Kairomones are the volatiles emitted by plants that benefit many predators by guiding them to prey or potential host insects.

Synomones ("with" or "together") are compounds that are beneficial to both the receiver and the sender such as volatiles emitted by flowers that attract bees for pollination.

The practical goal of semiochemical research is to develop techniques and methods for insects' pest control. Semiochemical research is placed in Pasteur's Quadrant of the Stokes model. It is based on the research in fine synthetic organic chemistry, but the final goal is still to develop solutions for agricultural problems, insects' pest population control, through applied research in the experimental field.

Since 1880s scientists used female insects to lure males into traps. Since the 1950s up until today, more than 3000 semiochemicals connected to the chemical communication of insects have been identified. Research on semiochemicals involves continued molecular mapping, synthesis, and studies of biosynthesis. Biologist and entomologist try to understand the neurophysiological sensory functions of insects and how hormonal regulation in insects affects pheromone biosynthesis and release.

Synthetic pheromones represent a new breeding prevention method for crop pest control. In sustainable agriculture using pheromones to control pests could drastically reduce the use of pesticides. The idea is to use an artificially synthesized scent, synthetic pheromones, to "attract and kill" into a trap the pests or to disrupt mating communication between male and female pests, thus preventing them from mating and lowering the population density of the next generation of the pests. These pheromones are specific and selective, have no effect on beneficial insects, such as pests' natural enemies or on other living organisms. Synthetic pheromones mimic the natural pheromones. Fascinating and somehow ironic is that the substances involved in perpetuation of the insects species can be used to control insect pest.

"Pheromonists," chemist researchers' team from "Raluca Ripan" Institute for Research in Chemistry Cluj – Napoca, Romania, is working to develop a variety of organic synthetic insects' pheromones and with multidisciplinary teams, biologists, entomologists, agronomists as partners in projects, develop new IPM friendly environmental techniques and technologies for insects' pest control.

Synthesis of pheromones and proposed IPM environmentally compatible strategies as monitoring or mass trapping of some coleopterean species (beetles), with the aim to reduce pest populations under economic damage thresholds are presented below.

The overall objective of research presented in this chapter is to find a technique using pheromones for protect : (1) maize crop against the West Corn Rootworm (WCR), *Diabrotica virgifera virgifera* LeConte (Coleoptera, Chrisomelidae); (2) potato crops against Colorado potato beetle (CPB), *Lepidoptera decemlineata* Say (Coleoptera, Chrisomelidae); (3) coniferous forests against six-spined spruce bark beetle, *Pityogenes chalcographus* (Coleoptera, Scolytidae).

2. Colorado potato beetle aggregation pheromones in IPM techniques for potato plant protection

2.1. Chemical ecology of Colorado potato beetle

The CPB, *Leptinotarsa decemlineata* (Say), Coleoptera, Chrisomelidae, native from Mexico, identified as major pest of potato plants first time in America in 1824, arrived in Europe in 1922, via cargo ships during World War I and subsequently colonizing all of Europe except for the British Isles and Scandinavia. Then, CPB continued to expand into eastern Europe and then central Asia and western China [9].

CPB (*L. decemlineata*) attacks potatoes (*Solanum tuberosum* L.) and various other cultivated crops such as tomatoes (*Solanum lycopersicum*) and aubergines (*Solanum melongena*). It also attacks wild solanaceous plants, which occur widely and can act as a reservoir for infestation. The adults feed on the tubers of host plants in addition to the leaves, stems, and growing points [10, 11] (**Figure 1**). Both adults and larvae feed on foliage and may skeletonize the crop.

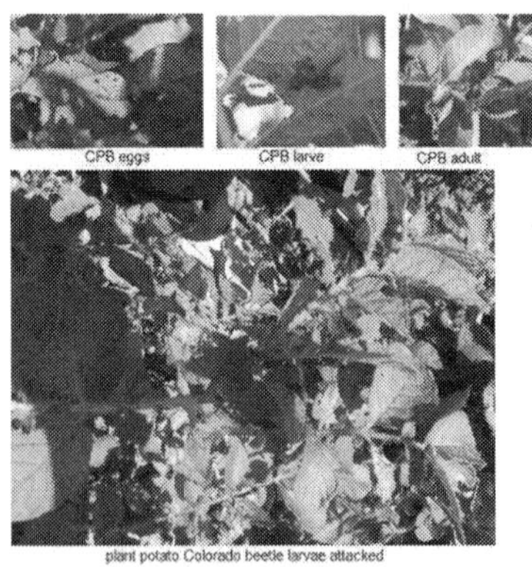

Figure 1. *Leptinotarsa decemlineata* (Say) adult on potato leaves (own photo on experimental plots).

CPB overwinter in the soil as an adult. The beetles become active in the spring. Females lay 800 of orange colored eggs in groups of two or several dozens for a period of 4–5 weeks. Larvae hatch after 4–9 days. Larval stage lasts 2–3 weeks and then the larvae hide in the ground. During their complete larval stage (3–4 weeks), CPB larvae consume approximately 40 cm^2 of potato leaves while adults can eat up to 10 cm^2/day [12]. It is well known in Europe, where the CPB population increased dramatically during and immediately following World War II and spread eastward.

Insecticides are currently the main method of beetle control on commercial farms. However, many chemicals are often unsuccessful when used against this pest because of the beetle's ability to rapidly develop insecticide resistance. The Colorado potato beetle has developed resistance to all major insecticide classes, although not every population is resistant to every chemical. The secret of Colorado potato beetle's success as a pest is its diverse and flexible life history coupled with a remarkable adaptability.

Now because of the inevitable decline of effective insecticide treatments, research should focus even more on the development of new control methods and approaches. Some methods such as cultivating GM plants are not seen positively by consumers, and farmers have abandoned them due to lack of buyers [13]. Researches for better understanding of insect's biology and lifestyle could permit entomologists and chemists to devise new control techniques.

The use of semiochemical attractants to improve insecticide treatments should be considered as an innovative approach of CPB management. Chewing insects are indeed more sensitive to volatile organic compounds (VOCs) released by their host plants because the damage they induce in plant tissues increases the release of these compounds [14].

It is necessary but very difficult to find the cocktail of natural odors within which the quantitative proportion of each compound is as close as possible to that of the naturally emitted blend [15]. The challenge consists in finding the appropriate molecules and their ratio, instead of trying to include as many compounds in the mixture as possible [14].

Researchers from Université de Liège and from ARS-USDA Beltsville, Maryland, USA review alternative strategies to control CPB populations [16]:

- biotechnological methods using intercropping cultures for disrupt the CPB adults perception of potato VOCs;
- trapping beetles using baits with synthetic mixtures of aggregation pheromone and/or volatiles kairomones;
- antifeedant sprays on potatoes;
- the potato plant recognizes the presence of CPB through chemical signals. By genetic manipulations increase the natural capacity of the plant to trigger defense mechanisms [16].

2.2. Experimental research using synthetically CPB aggregation pheromones

Our research is related to biotechnology that uses "chemical messengers" sending or receiving information for pest control in potato crops. Such "chemical mediators," which induce a certain behavior, are aggregation pheromones—intraspecific messengers and kairomones—interspecific messengers—chemical signals emitted by the host plants.A male-produced aggregation pheromone was identified for the Colorado potato beetle, *L. decemlineata* (Say) in 2002 by Dickens et al. [17] (**Figure 2**).

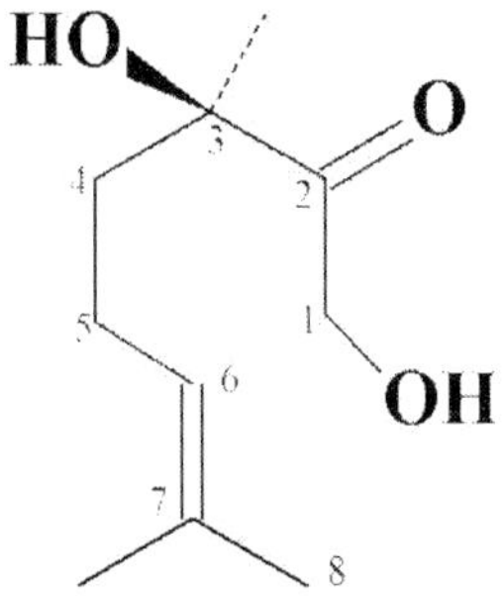

Figure 2. (*S*)-3,7-Dimethyl-2-oxo-6-octene-1,3-diol, aggregation pheromone of CPB [(*S*)-CPB].

The biological effect of (*S*)-CPB was first evaluated in a Y-olfactometer on male and female CPB, both of which were highly attracted. The (*R*)-enantiomer and the racemic mixture were not attractive [17]. CPB larvae also seem capable of perceiving the aggregation pheromone produced by adults, but further studies are needed to characterize larval behavior [18].

2.2.1. Synthesis of CPB aggregation pheromone

The synthetized (*S*)-CPB by route presented below (**Figures 3** and **4**) was analyzed and used in the field tests with extracts from potato plants (leaves) contain substances which function as kairmones.

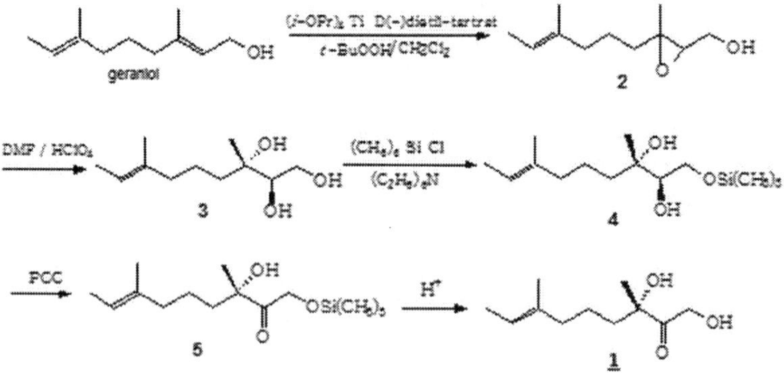

Figure 3. The way to prepare in ICCRR laboratory synthetic pheromone (*S*)-CPB.

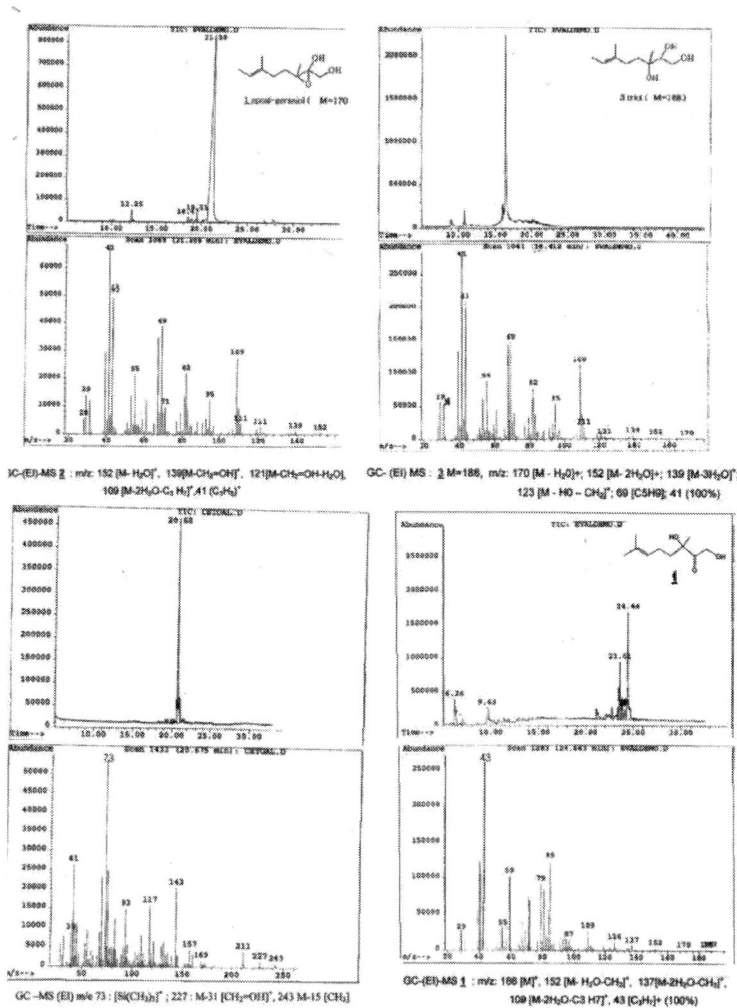

Figure 4. (EI) GC-MS analysis (Hewlett-Packard 5972 GC-MSD, capillary column HP-5MS (30 m × 0.25 mm × 0.25 μm) synthons and *S*-(CPB) aggregation pheromone.

2.2.2. Testing S-CPB attractivity for L. decemlineata: field experiments

Field experiments were conducted in three different locations from Transylvania area, Romania: Research-Development for Potato Station Targu-Secuiesc (**Figure 5**), University of Agricultural Sciences and Veterinary Medicine Cluj-Napoca, USAMV Research Station situated in Jucu-Cluj county, and Agricole Research-Development Station Turda (**Figure 6**).

Figure 5. Experimental plots from RDPS Targu-Secuiesc (own photo on experimental plots).

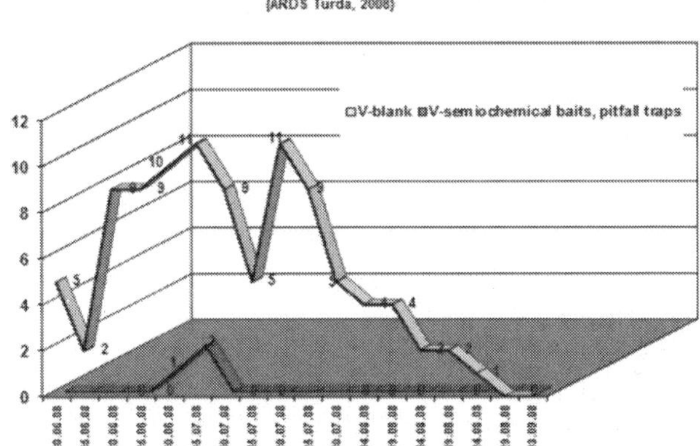

Figure 6. Adult CPB migration dynamic on potato plants in two plots (blank and experimental) ARDS Turda, Cluj, România.

Considering aggregation pheromone as attractant we try to get out the pest from potato field attract CPB a pitfall trap, a container with potato leaves alcoholic extract. Baits were placed on potato plants, close to the pitfall trap, "attract and kill" technique. The pheromonal baits are 0, 1 g *(S)-3,7-dimethyl-2-oxo-6-octen-1,3-ol* and *2-phenyl-ethan-1-ol* [(*S*)-CPB] impregnated on rubber stopper (**Figure 7**).

Experiments and observations from Research Development for Potato Station (RDPS) Targu-Secuiesc are presented below.

The experimental plots were artificially colonized with CPB. Each plot have 28 rows spacing 0.75 m and a distance between plants 0.30 m. Thirty CPB adults, collected from elsewhere, were placed on potato plants row nos. 13 and 14. The pitfall with kairomone (potato lives extract) and baits with pheromones are located on row no. 4 from the edge of the plot. In days 1, 3, 5, 7, and 10, CPB migration from row no. 14 ("start point") to row no. 4 "finish point" was observed. After day 10, 13 CPB adults were found on row no. 4, area with pheromonal baits.

Figure 7. Pheromonal baits and pitfall traps with kairomones.

The experiment conclusions are as follows: aggregation pheromone bait attracts beetle, but CPB did not reach into the trap, pitfall with kairomone is not efficient, probably the trap design is inadequate, to capture CPB must be used another type of trap such as a small wing trap (**Figure 20**).

2.2.3. Field experiments in USAMV Research Station situated in Jucu-Cluj county

In the field, the Colorado beetles were observed on potato plants. The experimental plots were located at 200 m distance from the blank plot. Pitfall traps and pheromone baits were placed in 40 m^2 area each, the experimental plot was at 20 m distance. A significantly number of CPB adults were identified by counting, crowded around traps, relative to the place where no pheromone traps were placed. The adults were aggregate, during egg lying, around traps in an area of about 18 m^2, with a circle radius of 2.5 m. This result shows that the behavior induced by this pheromone attract the beetles into the area, but these beetles do not try to touch the pheromone source, as occurs if the attractants are the sexual pheromone. Noteworthy, there was a higher concentration of adults in plots' edges and especially an affinity for plants infected with viruses.

2.2.4. Experimental research in the field of Agricola Research-Development Station Turda

At this location, the traps were placed on 0.8 ha potato plants' experimental plot, located at 20 m distance between them. Data placement of traps were made from June 5, following the evolution of both generations of the pest. Observations were made from June 10, and continued until September. Dynamic observation was performed each 5 days. Besides the abundance of adults, an observation on attack frequency (%) in each variant (experimental and blank) was performed. In 2008, the abundance of adults was lower compared to previous years, around 80 adults in the period between June 10 and late July, untreated version. In the experimental lot, the abundance was 19 adults in the mentioned period in the six traps.

Because the abundance of adults was lower this year, the frequency of attacks was insignificant. Thus, in the untreated lot, the attack rate was of about 40% and in the experimental lot it was 10%.

Pitfall with (*S*)-CPB as bait-attractant compositions was placed in the potato plant field. The effect of aggregation pheromone has been a migration, colonization beetles of both sexes, for "frontier" where they were installed traps and dispensers. In all the experiments, in the pitfall were found other insects of the same order as CPB, and many CPB adults were found around, but not in trap. The above results show a good aggregation capacity of the pheromone, but the traps still have to be perfection, because not all pests attracted by the pheromone are also captured.

By capturing their pest population, the results show a fall below economic threshold without affecting the potato crop or ecosystem.

It is necessary to continue to explore alternative control methods using semiochemicals and studying to better understand behaviors generated by these semiochemicals. The chemical ecology of CPB is not yet completely understood and this incomplete knowledge makes semiochemical-based approaches inefficient when compared to traditional insecticide treatments. The management strategies for CPB control must be flexible and adaptable to ever-changing circumstances [16].

3. Pheromones for maize crop protection

3.1. Chemical ecology of western corn rootworm (*D. virgifera virgifera* LeConte)

Pest description: Class: Insect; Order: Coleoptera; Family: Chrisomelidae; Genus: Diabrotica.

Western corn rootworm (WCR) *D. virgifera virgifera* LeConte infests corn crops in North America and since 1992 has been reported in Europe [19]. In Romania, WCR was first reported in 1996 at Nadlac (Arad County), near the Hungarian border and this quarantine pest migrates eastward.

The WCR beetles are about 5–7 mm long. Adults have a dark head, a yellow pronotum, and a yellow abdomen. The legs covered with short hairs are dark brown in males and brown in females. Male's body color is greenish-yellow and female's body has a yellow color [20] (**Figure 8**).

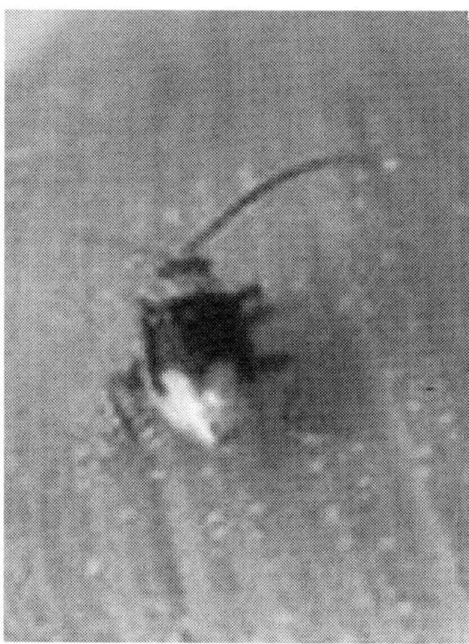

Figure 8. WCR (Diabrotica virgifera virgifera LeConte) adult (own photo on experimental plots).

The main damage is caused by WCV larva which lives in the soil and feed the roots and the adults feeding damage on corn silk and the maize in the milk stage, sometime on maize leaves or other species of plants from the spontaneous flora, but the multiplication of this species is assured by the maize crop [21] WCR (cucurbits, bean). Adults lay eggs in the soil and WCR larvae become active in May, attacking the roots of corn plants in development by drilling the cortical parenchyma, create tunnels in the central vascular tissue, which lead to the fall of the plant in windy day.

Factors that influence the pest propagation are as follows:

Soil: With good physical, chemical, and biological properties, loose and rich in humus, slightly acidic or alkaline, moist on top, favoring the breeding. Sandy soil is unfavorable for larvae, especially for the young during drought.

Climate: Gentle winter, with snow, free of strong winds; moistly spring; high air temperature for adult activity (until 30°C), favors the breeding.

Host plant: Larvae can feed on 22 species of plants, but they prefer maize and soybean. *Human intervention*—early seeding, the high density of the maize plants, excess nitrogen fertilization, irrigation, all these technologies contribute to the pest breeding.

Monitoring: Pheromone traps and yellow sticky traps were used.

Control: From nonpollutant methods it can be mentioned: pheromone traps, color traps, autochthonous natural enemies, and biological product Spinosad 240 SC (based on filamentous bacteria *Saccharopolyspora spinosa*) [22].

In the establishing of the pest's control strategy, an important part it has the prognosis of its appearance, which is based on the number of adults/plant (sticky traps with sexual pheromones baits), the number of larva and eggs/sample, the intensity of the caused damage to the silk of the corn cobs.

The reduction of the adult population is an important part in the reduction of the larva population of the next year and in the reduction of the damage to the cobs, which influences the production of beans and the quality of the seeds. It is recommended the control of adults, because their act by of destruction on the silk and implicitly compromise the pollination process when it is registered a density of more than 10 adults/plant at the commercial hybrids and five adults/plant at corn for seed [23].

3.2. Experimental research using synthetically sexual pheromone for WCR

Sexual pheromone of the *D. virgifera virgifera* Le Conte was identified by Guss et al., from virgin females of the WCR as 8-methyl-2-decanol propanoate (1,7-dimethyl-nonan-1-yl propanoate) [24].

The way of synthesis proposed and carried out in "Raluca Ripan" Institute for Research in Chemistry Laboratory is described in **Figure 9** and have five stages [25]. On this way, the racemic mixture of the four enantiomers was obtained [26].

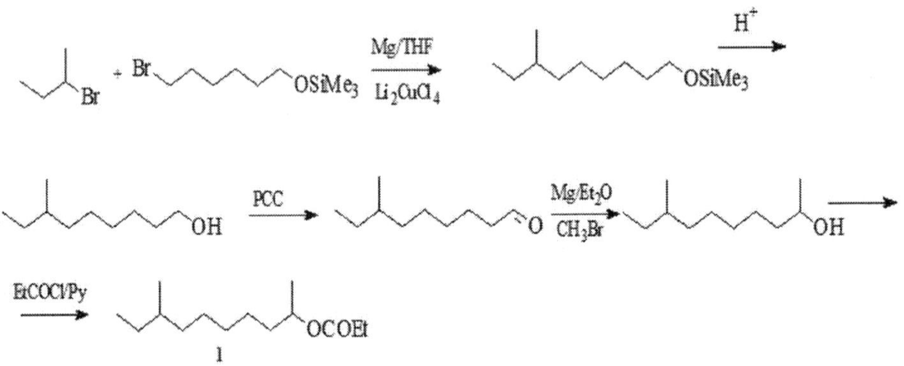

Figure 9. 1,7-Dimethyl-nonan-1-yl propanoate, WCR sexual pheromone synthesis.

The reaction yields for each stage was >74% and the intermediary and reaction products were identified through GC-MS, IR, and NMR [26].

In the case of the 1,7-dimethyl-nonan-1-yl propanoate there are four optically active forms or four optical isomers afferent to the two asymmetric carbon atoms C_1 and C_7. The four isomers, respectively: 1*R*, 7*R*; 1*S*, 7*S*; 1*R*, 7*S*; 1*S*, 7*R*; form two pairs of enantiomers (the **A** pair: 1*R*, 7*R*; 1*S*, 7*S*, the B pair: 1*R*, 7*S*; 1*S*, 7*R*) The GC-MS analysis carried out this time with a chiral column separates two pairs of diastereoisomers without separating each isomer (**Figure 10**, **Figure 11**).

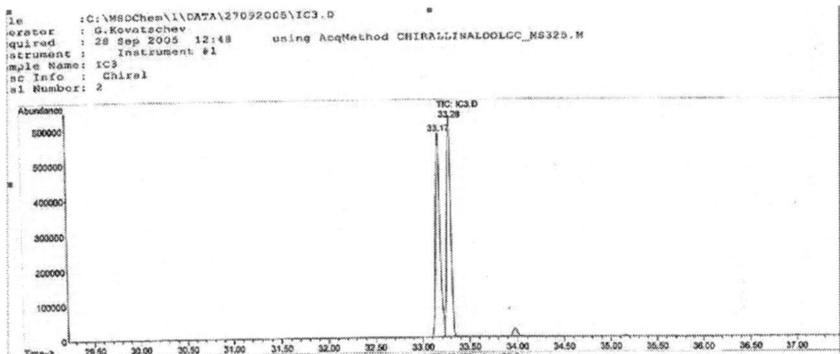

Figure 10. GC analysis of the 1,7-dimethyl-nonan-1-yl propanoate (GC Hewlett-Packard 5972 GC-MSD, capillary column HP-5MS (30 m × 0.25 mm × 0.25 μm).

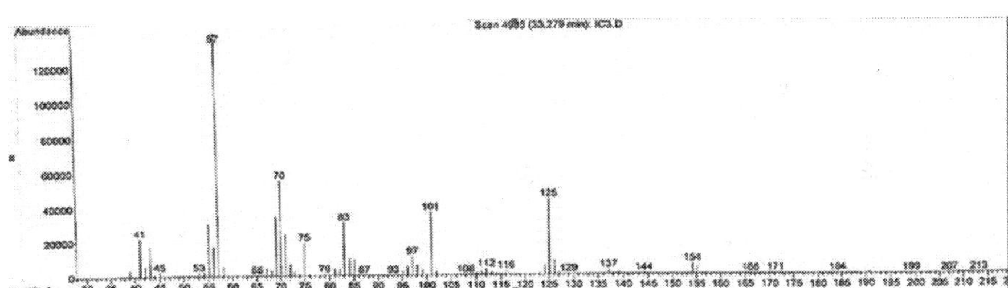

Figure 11. (EI)-MS spectrum of the 1,7-dimethyl-nonan-1-yl propanoate: GC-MS: tr = 13.91 min; *m*/*z*: 171 (M^+–COEt), 154 (M^+–OCOEt), 136, 125, 112, 101, 97, 83, 74, 70, 57 (100%), 43.

In the WCR, adults monitoring tests were carried out in the fields of Agricole Research and Developments Station (ARDS) Turda-Cluj county, Romania, the baits with racemic mixture using sticky traps showed a good attractivity.

The data from the graphic confirm both the existence of the attack of *D. virgifera virgifera* in the corn culture at the Turda Station, and the efficiency of the Romanian pheromone in comparison with others type of traps (**Figure 12**).

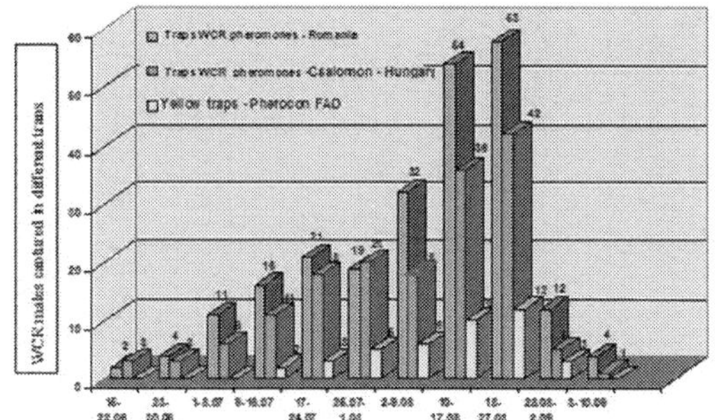

Figure 12. The efficiency of the Romanian pheromone in comparison with the imported one.

In 2005, at ARDS Turda, WCR was monitoring in field conditions, in two crop rotations: bean-wheat-corn and soy-wheat-corn, using sticky traps with bait sexual pheromones prepared in "Raluca Ripan" Institute for Research in Chemistry Cluj-Napoca.

The observations were carried out between July 12 and September 7, the number of WCR adults in this period being quite high: 921 WCR adults in the soy-wheat-corn crop rotation and 680 adults in the bean-wheat-corn crop rotation.

This pest can also develop in the soy culture, as shown in WCR adults numbers in the crop rotation with soy, as compared to the other one. The massive appearance of adults took place starting from the end of July and until the end of August, with a large number of adults in the second decade of August, when, in the traps with sexual pheromones were registered 100–248 adults/week.

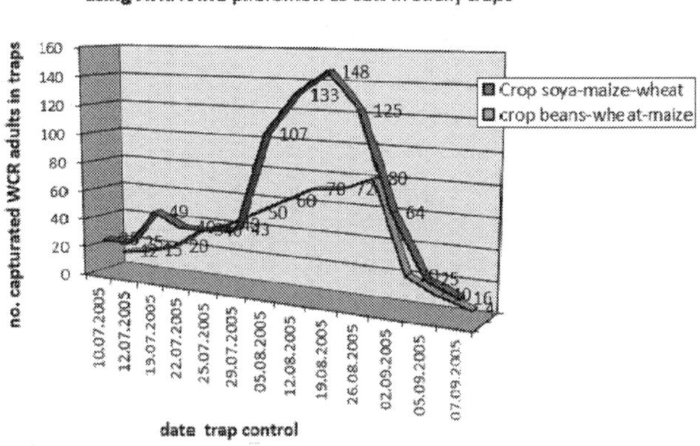

Figure 13. WCR males captured in two different crops.

The total number of adults of *D. virgifera virgifera* captured in the traps with pheromones, in the mentioned period, was of 1601 (**Figure 13**).

In Romania, start with July 1996, western corn rootworm (*D. virgifera virgifera*) was quarantine pest. In Transylvania, adults were monitored in the corn fields, e.g., 2002–2005 in ARDS Turda (**Figure 14**).

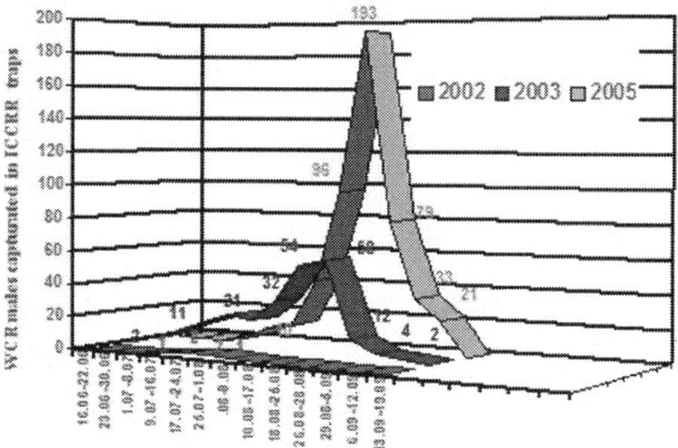

Figure 14. WCR pest "fly" dynamic in ARDS Turda fields (2002–2005).

As bait in sticky traps are 1,7-dimethyl-nonyl propanoate (WCR sexual pheromones) synthetized in "Raluca Ripan" Institute for Research in Chemistry (RRIRC) (**Figures 15** and **16**).

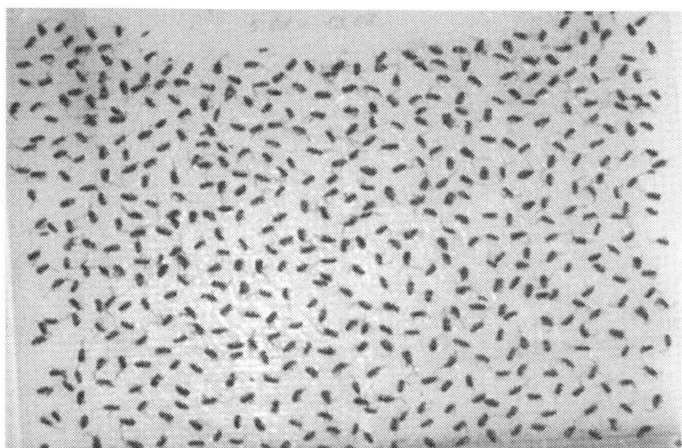

Figure 15. One single RRIRC sticky trap bait WCR sexual pheromones after 2 weeks exposed in corn field, ARDS Turda, Romania, 2005 (own photo on experimental plots).

Figure 16. RRIRC sticky traps (own photo on experimental plots).

3.3. Recommended biotech corn crop protection using pheromone traps

Together with other measures to reduce *D. virgifera virgifera* larval population (crop rotation, the seeding, and seed treatments), using synthetic sex pheromone traps is a recommended option for the pest management.

Traps used for western corn rootworm *Diabrotica virgifera virgifera* are placed on a stick in the ground or in more vigorous corn plant at the height of 1 m, usually 1–2 weeks before presumptive appearance of the pest (in June).

Adults begin to emerge, usually in late June, when the corn plant already has silk (the beetle's favorite food). If used for monitoring, 9 traps/ha have to be installed at the edge of the maize lot. The traps are inspected twice a week when count the captured beetles and clean the trap by removing butterflies, insects or leaves that accidentally entered on the sticky surface. Sticky plate is replaced twice per month, pheromone bait once per month, and observations are made in 5–7 days. Depending on the number of adults captured, the chemical treatment is indicated or not. So, monitoring and treatments recommended are as follows:

- If the number of catches is 5–8 WCR adults/trap in next year. If corn is sown on the same plots, the roots will be attacked by larvae. It is necessary either seed treated with insecticide or treatment ground for larvae. Treatment with granular soil insecticide is done either when seeding or at first diggings. If there are larvae (i.e., eggs deposited in the previous year), only ground treatment is insufficient.
- If the number of capture is 10 adults/trap—corn for consumption or five adults/trap—corn for sowing, treatment is required for adults.

The field tests show how important are traps with pheromonal baits for monitoring the appearance of WCR adults in crops and for decreased adult populations during mating season so that generations of larvae in the next year are reduced.

4. Aggregation pheromones used in pine forest protection

4.1. Chemical ecology of six-spined spruce bark beetles *P. chalcographus* L.

Pest description: Class: insect; Order: Coleoptera; Family: Scolytidae; Genus: Pityogenes.

P. chalcographus infests Norway spruce [*Picea abies*], especially the younger trees or, in competition with another bark beetles *Ips typographus*, the upper regions of older trees. Bark beetles (Coleoptera, Scolytidae) must compete for food and space in which to reproduce within the relatively thin phloem layer of their host tree [27]. *P. chalcographus* is rather small for bark beetles, being only 2 mm long and weighing 1.2 g, the color is dark brown almost black [28].

Biology: In the Nordic countries, *P. chalcographus* has a single generation in a year. Romania had two flights per year. The first flight was in April-June (76–94% from the flight on all growing season), the second flight was in July-August (6–24% from the flight on all growing season) [29].

Damage: Both sexes are aggregated through male pheromone released. The attracted males want to join the attack and secure an area for his and several female's young. The female deposits their eggs in galleries excavated in the vascular cambium and secondary phloem. The phloem layer is only about 2–4 mm thick and rich in nutrients; successful breeding is dependent on the death of these tissues. Larval galleries have a length of 2–4 cm, are dense, well printed on bark and wood weak [30]. Aggressive bark beetle species like *P. chalcographus* are associated with pathogenic blue stain fungi, which help them to overcome the defense reaction of the host tree. Some of the fungi in this group are pathogenic and may play an important role in the death of the tree by blocking water conduction [31] or indirectly by overstimulating tree defense mechanisms that may exhaust the host [32, 33].

Host: Norway spruce is the preferred but not the only host. Bark beetles belong to the family of ypids. Usually, these insects are attracted to and breed on trees felled or broken by wind, on trees affected by fire or sunstroke, on trees severely debilitated by drought or pollution, and the trees having lost defense capability. Most of the time, these beetles live under the bark of trees, feeding inner part of the bark, and leave this place just to seek new sources of food.

Monitoring: At the temperature of 16.8–17°C, beetles *P. chalcographus* become active and could be monitored [34].

Control: Pest control is carried out by pheromone traps. The use of pheromones by "mass trapping" technique is one of the few ways to protect the forest ecosystem.

4.2. Experimental research using synthetically aggregation pheromone for bark beetle *P. chalcographus*

4.2.1. Chemical synthesis of the main component

The bark beetle *P. chalcographus* aggregation pheromone is a "cocktail" with four components: 2-ethyl-1,6-dioxaspiro-[4,4]-nonane (Chalcogran)—the main component [35] and secondary components: methyl-*E*2,*Z*4-decadienoate; α-pinene and ipsdienol.

The proposed synthesis *for 2-ethyl-1,6-dioxaspiro-*[4,4]-*nonanne* is represented in (**Figure 17**, **Figure 18**), it has four steps with 1,6-hexanediol as starting substance. It is an original reaction path except for the last stage, where the reaction conditions used by Cekovic and Bosnjak [36] at the cyclization of 1,7-nonanediol were modified [37, 38].

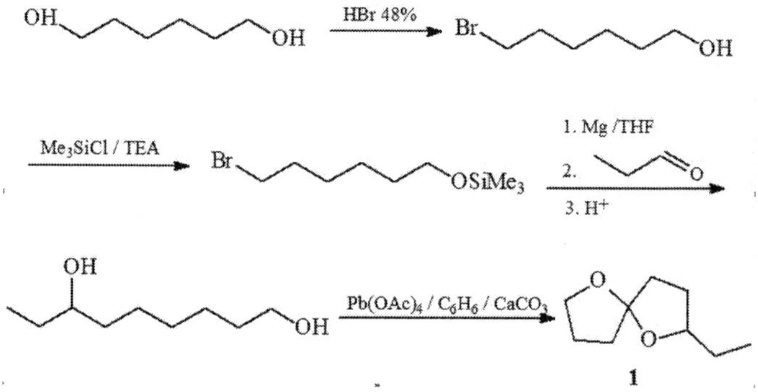

Figure 17. The reaction path which was proposed and carried out in laboratory.

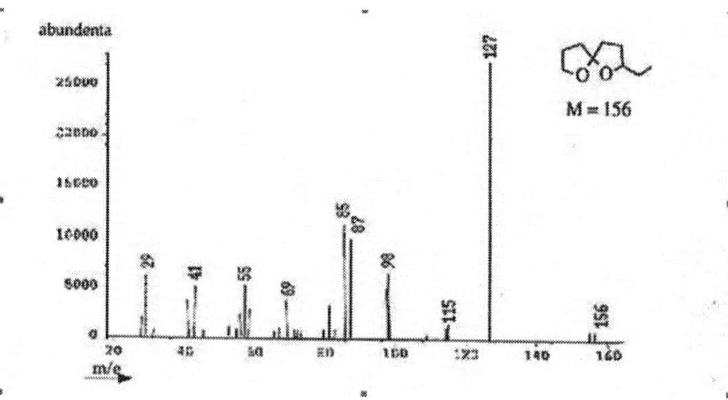

Figure 18. The mass spectra of the main product of reaction: 2-ethyl-1,6-dioxaspiro-[4,4]-nonane (**1**).

http://dx.doi.org/10.5772/64278

The 2-ethyl-1,6-dioxaspiro-[4,4]-nonane (**1**), having two asymmetric centers, has two pairs of optical diastereoisomers: **A** (2*R*,5*S*-**1**; 2*R*,5R-**1**) and **B** (2*S*,5*S*-**1**; 2*S*,5*R*-**1**) (**Figure 19**). The natural pheromone contains the pair of diastereoisomers (2*S*,5*R*)-**1**—biologically active and (2*S*,5*S*)-**1** —biologically inactive [39, 40].

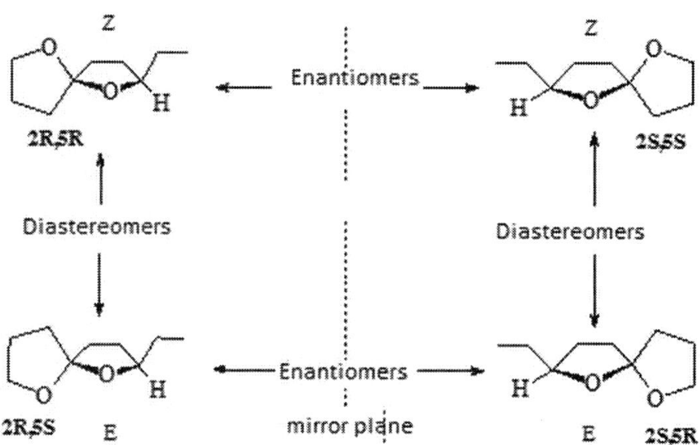

Figure 19. Enantiomers and diastereomers for 2-ethyl-1,6-dioxaspiro-[4,4]-nonane.

All the four stereoisomers were obtained by synthesis and this racemic was tested.

4.2.2. Field experiments

The biological activity, respectively the efficiency of the Romanian pheromone baits is tested in comparison with other imported compound. All the tests were obtained in Brasov area (Romania) between 2001 and 2004 (**Tables 1** and **2**).

Year	Location of the experiments Romania, Brasov	The type of trap	The type of bait	No. of traps	The period and duration of observation (days)	The number of captures	The intensity of the attraction
2001	Gârcin	Wing	Atrachalc	3	16.05–12.07	1.405	8.07
	(OS Săcele)		Baits import	3	(58 days)	70	0.4
2004	Tamina	Theysohn	Atrachalc	3	31.05–20.07	1.292	8.4
	(OS Braşov)		Baits import	2	(51 days)	503	4.9

Table 1. Experimental results in Romanian forests, 2001–2004.

Year	Location of the experiments	The type of Trap	The type of bait	No. of traps	The period and duration of observation (days)	The number of captures	The intensity of the attraction
2002	Gârcin (OS Săcele)	Wing	Atrachalc	5	21.05–28.07 (68 days)	35.531	104.5
2003	Gârcin (OS Săcele)	Wing	Atrachalc	3	19.05–1.07 (43 days)	35.679	276.5

Table 2. Experimental results , Romanian forests, 2002–2003.

Figure 20. *Pityogenes chalcographus* wing trap installed in forest (own photo on experimental plots).

4.2.3. Conclusions

(1) In 2001 and 2004, tests show the increased attractiveness of baits with the Romanian pheromone—Atrachalc, compared to another baits (import), irrespective of the type of trap used.

(2) In 2002 and 2003, the tests carried out with the Atrachalc baits and wing traps show in the same location different no captured beetles according to the period of time when the traps were placed and the observations were made. Besides the time factor, weather conditions or other elements from the ecosystem could influence catches

(3) 2009 comparative tests in Brasov area with different lures obtained higher level of the captured beetles using Atrachalc—Romanian baits. Wing traps with Atrachalc lure are recommended by experts for Romanian forests [41].

4.3. Recommended biotechnique: *P. chalcographus* "mass-trapping"

For monitoring and control bark beetles, pheromone lures are placed in wing-type trap (**Figure 20**). The traps are placed at the forest edge at about 5 m in the case of old forests and about 15 m in the case of young forests. Between the traps, distance is 30–50 m. To a high infestation, 2–3 traps/ha are used and for low infested forest one trap/ha is used. Traps are installed in late April to late August usual on a tree already attacked by bark beetles. Pheromone baits are replaced at no more than 6 weeks.

5. Conclusions

A changing climate with higher growing season temperatures and altered rainfall patterns make control of native and invasive insects an increasingly urgent challenge. Treatments with increasing amounts of insecticides are not a solution; it is time to intensify interdisciplinary research on semiochemicals based on a scientifically sound understanding of pest biology to provide the urgently needed and cost-effective technical solutions for sustainable insect management worldwide [42].

Using pheromones in order to protect the above-mentioned crops is first of all an ecofriendly method, avoiding these ways the overloading environment with insecticides. The ecosystem remains unaffected due to the high selectivity and specificity of the semiochemicals. Sex pheromone baits in sticky traps attract insects very selective, only the species that emitted for mating the natural sex pheromons.

This method does not affect another components of the ecosystem such as soil, air, water, or animals. The insect pest population falls below the economic damage threshold. In the case of the bark beetle is the most efficient combating method, because this pest lives (acts) underneath the bark, a place where insecticides cannot be applied.

The same happen in the case of WCR because it is very difficult to use pesticides in the maize crop. Because Colorado potato beetles develop rapidly resistance to insecticides using aggregation pheromone as bait in a proper trap could be a solution to control this pest.

All studies and experimental reviews above-mentioned have enhanced knowledge of chemical communication in and highlight the potential of semiochemicals as a component of future integrated management strategies.

Acknowledgements

The original researches were supported by PN II grants financed by MEN-ANCS CeEx—BIOTECH 116, INVENT 160, grant CNCSIS-MEN Romania. We thank Dr Felicia Muresanu—Agricola Research Development Station Turda, Prof. Dr Gavrila Morar—University of

Agricultural Sciences and Veterinary Medicine Cluj, Dr Daniela Popa—Research-Development for Potato Station Targu-Secuiesc for testing in the fields ICCRR-synthetic Pheromones.

Author details

Maria Pojar-Fenesan* and Ana Balea

*Address all correspondence to: maria.fenesan@ubbcluj.ro

Babes-Bolyai University, "Raluca Ripan" Institute for Research in Chemistry, Cluj-Napoca, Romania

References

[1] Allen P, Dusen DV, Lundy J, Gliessman S. Expanding the definition of sustainable agriculture. Sustainability in the Balance. 1991;3:1–8.

[2] Briejer CJ. The growing resistance of insect to insecticides. Atlantic Naturalist. 1958;13(3):149–155.

[3] Smith RF, Hagen KS. Impact of commercial insecticide treatments. HILGARDIA. Journal of Agricultural Science, California University. 1959;29(2):81–96.

[4] Stern VM, van den Bosch R. Field experiments on the effects of insecticides, HILGARDIA. Journal of Agricultural Science, California University. 1959;29(2):1–80

[5] Gill HK, Garg H. Pesticide: environmental impacts and management strategies. In: Solenski S, Larramenday ML, editors. Pesticides—Toxic Effects. Rijeka, Croatia: Intech; 2004. pp. 187–230.

[6] Edward M. Barrows,. Animal Behaviour Desk Reference: A Dictionary of Animal Behaviour, Ecology, and Evolution, 2rd ed., CRC Press; Boca Raton London, New York Washington D.C., 2000.

[7] Norin T. Semiochemicals for insect pest management. Pure and Applied Chemistry. 2007;79(12):2129–2136. DOI: 10.1351/pac200779122129. © 2007 IUPAC.

[8] Karlson P, Butenandt A. Pheromones (ectohormones) in insects. Annual Review of Entomology. 1959;4:39–58. DOI: 10.1146/annurev.en.04.010159.000351.

[9] Weber DC. Colorado beetle: pest on the move. Pesticide Outlook. 2003;14:256–259.

[10] Hare JD. Impact of defoliation by the Colorado potato beetle on potato yields. Journal of Economic Entomology. 1980;7:369–373.

[11] Hare JD. Ecology and management of the Colorado potato beetle. Annual Review Entomology. 1990;35:81–100.

[12] Ferro DN, Logan JA, Voss RH, Elkinton JS. Colorado potato beetle (Coleoptera: Chrysomelidae) temperature-dependent growth and feeding rates. Environmental Entomology. 1985;14:343–348.

[13] Guenthner JF. Consumer acceptance of genetically modified potatoes. American Journal of Potato Research. 2002;7:309–316.

[14] Szendrei Z, Rodriguez-Saona C. A meta-analysis of insect pest behavioral manipulation with plant volatiles. Entomologia Experimentalis et Applicata. 2010;134:201–210.

[15] Visser JH, Ave DA. General green leaf volatiles in the olfactory orientation of the Colorado beetle, *Leptinotarsa decemlineata.* Entomologia Experimentalis et Applicata. 1978;24:738–749.

[16] Sablon L, Dickens JC, Haubruge E, Verheggen FJ. Chemical ecology of the Colorado potato beetle, *Leptinotarsa decemlineata* (Say) (Coleoptera: Chrysomelidae), and potential for alternative control methods. Insects. 2013;4:31–54. DOI: 10.3390/insects4010031.

[17] Dickens JC, Oliver JE., Hollister B, Davis JC, Klun JA. Breaking a paradigm: male-produced aggregation pheromone for the Colorado potato beetle. Journal of Experimental Biology. 2002;205:1925–1933.

[18] Hammock JA, Vinyard B, Dickens JC. Response to host plant odors and aggregation pheromone by larvae of the Colorado potato beetle on a servosphere. Arthropod-Plant Interaction. 2007;1:27–35.

[19] Berger H. Report about the first meeting on *Diabrotica virgifera virgifera* Le Conte. In Graz IWGO. Newsletter. 1995;15(2):24.

[20] Florian T, Oltean I, Bunescu H, Florian V, Varga M. Research regarding the external morphology of *Diabrotica virgifera virgifera* Le Conte, western corn root worm. ProEnvironment. 2011;4:233–236. ISSN: 2066-1363.

[21] Bača F. New member of the harmful entomofauna of Yugoslavia, Diabrotica virgifera virgifera Le Conte (Coleoptera, Chrysomelidae). Zastita Bilja, 1994, 45, 125 – 131.

[22] Grozea I, Badea A-M, Chiriță R, Carabeț A, Adam F. *Strategies to combat invasive species Diabrotica virgifera,* Scientific Papers. Seria Agronomie. 2007;50:496–500, Ion Ionescu de la Brad, Iaşi, ISSN 1454–7414 (CABI), http://www.revagrois.ro/cuprins.php.

[23] Sivcev I, Manojlovic B, Krojajic S, Dimic N, Draganic M., Distribution and harmfulness of Diabrotica virgifera LeConte (Coleoptera: Chrysomelidae), new pests of corn in Yugoslavia. Plant protection journal. 1994. 45(1):19–26.

[24] Guss PL, Tumlinson JH, Sonnet PE, Proveaux AT. Identification of a female-produced sex pheromone of the western corn rootworm. Journal of Chemical Ecology. 1982:8:545–556. DOI: 10.1007/BF00987802.

[25] Pojar-Fenesan M, Pop L, Oprean I, Balea A. Synthesis of racemic 8-methyl-2-decyl propanoate, the sex pheromone of western corn rootworm *Diabrotica virgifera virgifera* LeConte (Cleoptera, Chrisomelidae). Revue Roumaine de Chimie. 2001;45(3):263–266.

[26] Pop LM, Pojar-Feneşan M, Balea A, Oprean I. Process for obtaining racemic 1,7-dimethyl-nonyl propanoate, sex pheromone of corn root worm Diabrotica virgifera virgifera LeConte (Coleoptera, Chrisomelidae). RO Patent No. 119 362; 2007.

[27] Byers JA. Avoidance of competition by spruce bark beetles, *Ips typographus* and *Pityogenes chalcographus*. Experientia. 1993;49:272–275. Birkhauser Verlag, Basel/ Switzerland.

[28] Kolk A, Starzyk JR. Smaller European spruce bark beetle (*Pityogenes chalcographus*). The Atlas of Forest Insect Pests. The Polish Forest Research Institute. Multico Warszawa; 1996. 705 pp.

[29] Fora CG, Lauer KF, Damianov S. Research regarding the biology of the pest *Pityogenes chalcographus* L. (Coleoptera, Scolytidae) in the Nădrag-Padeş area (Timiş County). Research Journal of Agricultural Science. 2007;39:431–438.

[30] Simionescu A, Mihalciuc V, Chira D, Lupu D, Vlăduleasa A, Vişoiu D, Rang C, Mihai D, Mihalache G, Ciornei C, Olenici N, Neţoiu C, Iliescu M, Chira F, Tăut I. Protecţia pădurilor. Editura Muşatinii, Suceava; 2000. 867 p.

[31] Paine TD, Raffa KF, Harrington TC. Interactions among scolytid bark beetles, their associated fungi, and live host conifers. Annual Review of Entomology. 1997;42:176–206.

[32] Kirisits T, Diminic D, Hrasovec B, Pernek M, Baric B. First reports of silver fir blue staining. Ophiostomatoid fungi associated with *Pityokteines spinidens*. Croatian Journal of Forest Engineering. 2009;30:51–57.

[33] Lieutier F, Yart A, Salle A. Stimulation of tree defenses by Ophiostomatoid fungi can explain attack success of bark beetles on conifers. Annals of Forest Science. 2009;66:22.

[34] Lobinger G. The air temperature as a limiting factor for the two spruce bark beetles, Ips typographus L. and Pityogenes chalcographus L. (Col., Scolytidae). Anz Schädlingskde, Pflanzenschutz, Umweltschutz. 1994;67:14–18.

[35] Francke W, Heemann V, Gerken B, Renwick JAA, Vite JP. 2-Ethyl-1,6-dioxaspiro[4,4]nonane, principal aggregation pheromone of *Pityogenes chalcographus* (L.). Naturwissenschaften. 1977;64:590–591.

[36] Cekovic Z, Bosnjak. Synthesis of spiro-ketal pheromones. Journal of Croatica Chimica Acta. 1985;58:671.

[37] Balea A, Pojar-Fenesan M, Pop L, Oprean I, Ciupe H. Synthesis of racemic 2-ethyl-1,6-dioxaspiro-[4,4]-nonane, the aggregation pheromone of *Pityogenes chalcographus*

(Coleoptere, Scolitidae). Revue Roumaine de Chimie. 2003;34:465–469. DOI: 10.1002/chin.200330240.

[38] Pop L, Pojar-Fenesan M, Balea A, Oprean I. Procedeu de preparare a racemicului 2-etil-1, 6-dioxaspiro-[4,4]-nonanului, feromon de agregare al gandacului de scoarta *Pityogenes Chalcographus* (Coleoptere, Scolitidae). RO-Patent: 120339 B1; 2005.

[39] Schurig V, Weber R. Use of glass and fused silica open tubular columns for the separation of structural, configurational and optical isomers by selective complexation gas chromatography. Journal Chromatography. 1984;289:321–332. DOI: 10.1016/S0021-9673(00)95097-0.

[40] Byers JA, Hogberg H-E, Unelius CR, Birgersson G, Lofqvist J. Structure-activity studies on aggregation pheromone components of *Pityogenes chalcographus* (Coleoptera: Scolytidae) all stereoisomers of chalcogran and methyl 2,4-decadienoate. Journal of Chemical Ecology. 1989;15:685–695. ISSN: 0098-0331 (Print) 1573–1561 (Online).

[41] Isaia G, Paraschiv M. Research concerning the effect of synthetic pheromones on *Pityogenes Chalcographus* L. in Braşov County. Bulletin of the Transylvania University of Braşov Series II: Forestry. Wood Industry. Agricultural Food Engineering. 2011;4:53.

[42] Witzgall P, Kirsch P, Cork A. Sex pheromones and their impact on pest management. Journal of Chemical Ecology. 2010; (1):80–100. DOI: 10.1007/s10886-009-9737-y.

Chapter 7

Detergents and Soaps as Tools for IPM in Agriculture

Tomislav Curkovic S.

Additional information is available at the end of the chapter

http://dx.doi.org/10.5772/64343

Abstract

This chapter presents extensive and updated knowledge from scientific and technical reports on the management of agriculture pests using detergents and soaps (D + S), with emphasis on their utility in integrated pest management (IPM) schemes. It includes a review on their environmental, ecological, and toxicological impacts, and their possibilities to become important tools for pest control, especially for those D + S having minimum risk, considering both current and newer products. The present knowledge of their modes of action on arthropods is addressed, revealing the need to better identify the mechanisms to optimize their use against crop pests. Their disadvantages are also analyzed, mainly the lack of residual effect and the potential toxicity to plants. Some ways these problems have been overcome are presented. A comparison of the direct costs of the use of conventional pesticides versus D + S, achieving statistically similar levels of control, is discussed, and scenarios where detergents are competitive (representing lower costs) are presented. There is also a review of the type of compounds reported in the specific literature, which leads to highlight the opportunities to develop agriculture detergents and soaps suited to local agriculture needs. New findings on D + S as co-adjuvants for conventional and biological pesticides, and their potential utilization as safe postharvest treatments against pest, are also presented. Finally, the authorization for soaps and detergents is also discussed, highlighting the need for a joint effort (state agencies, producers, researchers, etc.), in order to increase the offer and the use of detergents and soaps, partially replacing conventional pesticides, to take advantage of their potential as sustainable pest management tools, particularly for IPM programs, but also for organic and conventional productive schemes.

Keywords: detergents, IPM, soaps, surfactants, sustainability, toxicity

1. Introduction

1.1. Chemical control in integrated pest management (IPM) programs and detergents and soaps

Integrated pest management is a strategy developed to control agricultural pests and, at the same time solve problems derived from the extensive and intensive implementation of chemical control in conventional agriculture, where broad spectrum, specific action site, and persistent pesticides are used. Compounds with this profile have been called "conventional pesticides" and are responsible for causing resistance in pest populations, destruction of beneficial arthropods, and presence of pesticide residues in foods, soils, water, and air [1]. In order to obtain an economic, environmental, and ecologically sustainable food production, IPM encompasses several components, including cultural, biological, and chemical control [2]. Therefore, the use of pesticides is not excluded from IPM programs, for instance, when there is no other available tools to avoid economic damage [3], synergy occurs between chemical and biological control [4], or a diverse pest complex affects the crop [5]. Under those circumstances, the products used should target several sites and mechanisms (multisite), be shortly persistent in the environment and crops (non-residual), and have both a narrow spectrum (selective) and low toxicity to mammals. Many compounds having these attributes have been called "alternative pesticides", including oils, pheromones, botanicals, entomopathogens, and soaps and detergents, among the most frequently used [6, 7]. For definition purposes, agriculture detergents and soaps, from now on "D + S", correspond to surfactants from either natural or synthetic origin, formulated specifically for pest control or other uses in crops. Within these options, D + S have additional particularities, being relatively inexpensive, easy to produce and apply, versatile (controlling juvenile and adults), allowed as postharvest treatment, etc. [8, 9].

1.1.1. Resistance management

Resistance is a consequence of the elimination of susceptible genotypes and selection, over time, of the tolerant part of the population by the frequent and wide use of pesticides with specific sites of action that lose afterwards their capability to control pests [1]. The alternating use of conventional products with different action sites has been one way to face resistance, but a more holistic approach is necessary to provide a sustainable solution [10]. That is why IPM was developed during the second half of the twentieth century, attempting to either avoid or reverse resistance by replacing chemical control by other strategies, and/or by using several different chemicals with multiple modes of action, as D + S that, therefore, should become useful tools for IPM [8, 11].

1.1.2. Environmental, ecological, and toxicological issues

Environmental contamination, diversity threatening, and toxic effects on mammals and other animal species are well known and severe impacts from the use of conventional pesticides. Environmental toxicity by soaps, on the other hand, is considered very low [12], but detergents in wastewater (sometimes in large concentrations) are considered important pollutants when they reach rivers and streams, where they form foam layers and affect the aquatic fauna. However, the greater biodegradability of current surfactants has significantly reduced those problems [13]. Besides, sprays in farms should not massively reach water courses, therefore minimizing the potential impact in surface and groundwater. Based on studies of wastewater used for irrigation [14], some surfactants alter physical, chemical, and biological properties of some types of soils [15]. However, linear alkylbenzene sulfonates (LAS, widely used in detergents) are considered not to be a threat to terrestrial ecosystems on a long-term basis because of biodegradation [16], although nonylphenol has been questioned [17]. Thus, their impact depends largely on the type of surfactant chemistry, providing room for testing, selecting, and using those less hazardous products.

In general, D + S have low acute toxicity [18], particularly non-ionic or anionic detergents, which are, by far, less dangerous than conventional insecticides [19]. For instance, the soap Safer has an oral LD_{50} of 16.500 ppm (= median lethal dose, i.e., the amount of active substance per body weight required to kill half of an exposed population), which is by far less dangerous than conventional insecticides, including botanicals [12]. The risk should be even lower considering both the necessary dilution and the small chance of ingestion. Conventional pesticides on the foliage are an important risk for applicators by dermal exposure, making necessary reentry intervals after their application, which are not needed when D + S are used. Detergents can cause dermal [20] or eye irritation, but in general this type of exposure represents a very low risk to agriculture workers wearing the basic personal protective equipment, although some respiratory disorders have been reported to detergent exposure, mainly on asthma sufferers [21, 22]. There are some concerns regarding specific housecleaning products (e.g., those containing alkyl phenols), which have been related to breast cancer [23], although under normal exposure in the field the risks are reduced, since no systemic toxicity is expected for most D + S and several components of their formulations [18, 19, 24], but this issue needs a case-by-case analysis. Another important issue is the persistence of conventional pesticide residues in/on the marketable part of the crop that makes necessary to establish regulations of MRLs (maximum residue limits) for foods. Thus, PHIs (preharvest intervals) are established to comply with the law, whereas most D + S are not subjected to this type of restrictions. In fact, some D + S are applied right before harvest [9] and others are authorized for postharvest treatments [25], being easily washed off from the epidermis of fruits and vegetables by rinsing before consumption, having minimum risk and being therefore exempt of MRLs [26].

Regarding the impact on beneficial fauna in crops, D + S have been considered more selective than conventional insecticides, being compatible with biological control due to their low adverse impact on not sprayed insect and mites and the lack of residual activity [4, 27]. The only threat occurs by the direct spray or when the solution persists on the foliage, usually for short periods, killing predators and parasitoids. Therefore, the release of beneficial arthropods after a spray, once deposits are dry, allows them to survive. Available EIQ (environmental impact quotient that considers environmental and ecological threats) values for soaps indicate their low impact (e.g., 19.45 for M-Pede), close to most botanicals or IGRs (insecticide growth regulators), and smaller than those of horticulture oils [28]. However, no data on detergents were available. Therefore, research to identify efficient (current or new), but also nontoxic and ecologically safe D + S for pest control is required.

1.1.3. Legal and economic issues

Conventional pesticides are subject to a complex and expensive registration process where, after agronomic and toxicological reviews, they might obtain legal authorization to be used on crops. On the other hand, D + S are not necessarily subjected to registration, since some products are not labeled as pesticides, but as tree cleaners. However, it is important to transparent the real purpose of its use in agriculture [11]. Even when explicitly recommended to control pests, D + S should be easier to register after considering their risk assessment due to their low acute and chronic toxicity and, in some cases, their status as food additives or edible surfactants [29]. Considering the growing demand for residue-free foods, the eventual replacement of conventional pesticides for D + S will make those foods preferred by customers, increasing their value and making their trade easier. Therefore, all the actors involved should deeply assess D + S uses for pest control.

1.2. Modes of action of detergents and soaps as pesticides

The modes of action for D + S against pests have not been well understood yet [30, 31]. In fact, D + S are not considered on the IRAC (Insecticide Resistance Action Committee) lists that classify the pesticides mode of action for those with known specific target sites [32]. This is because D + S are not known to act at specific target sites, but at multiple sites [11]. Despite that, wax removal, arthropod dislodging, and drowning have been mentioned as lethal mechanism in D + S.

1.2.1. Wax removal

The arthropod epicuticle is mainly made of lipids. The outermost part is a wax layer constituted mostly by hydrocarbons, serving mainly for waterproofing to avoid dehydration [33]. This is a serious threat for small insects and mites, particularly those sessile and exposed individuals. It has been proposed that when arthropods are sprayed with detergent, lipids are removed from the epicuticle, losing its waterproof ability, which in turn causes important water losses and, finally, the death of treated pests [34]. In fact, a significant reduction in both residual epicuticular lipids and body weight (assumed to occur mainly due to water losses) on the obscure mealybug *Pseudococcus viburni* Signoret (Hemiptera: Pseudococcidae) sprayed with detergent solutions was measured ([35], **Table 1**). After the spray, water losses reached up to 3% of body weight 7 h after exposure, and residual waxes were 88–73% below when compared with the control (check) at 24 h. Mortality was positively related with both water losses and wax removal when the agriculture detergent TS 20135 was used, but no significant relationship was found when the surfactants alone (excluding the co-adjuvants from the formulation) were tested.

Santibáñez [35] proposed that mealybug mortality by exposure to detergents might be caused by several mechanisms, including the initial wax removal that might lead to further damage of the integument, but this was not demonstrated. Many reports of pest management with D + S reveal that individuals present a degreased and dehydrated aspect after exposure, suggesting that water losses might be involved in mortality. For instance, the cotton aphid *Aphis gossypii* Glover (Hemiptera: Aphididae) nymphs and adults were strongly dehydrated and their bodies collapsed when evaluated 48 h after the spray with an agricultural detergent [9]. Wax removal (assumed to lead to dehydration) is also evident after exposure to detergents, causing dramatic changes in mealybugs, even a few minutes after the spray ([8, 11], **Figure 1** shows effects on hemipterans either sprayed or immersed in solutions).

Treatments[1]	Detergent (mL a.i.[2]/100 mL)	Water loss[3] (mg)	Residual waxes[4] (mg/mL)
LC_{90}	8.17	1.85 a[5]	14.95 b[5]
LC_{50}	4.45	1.48 b	6.85 b
LC_{10}	0.74	0.89 c	54.76 a
Control	0.00	0.47 c	55.06 a

[1] LC = lethal concentration estimated by Probit analysis; study conducted using a Potter tower, control sprayed with water.
[2] Active ingredient, the sum of surfactants formulated in TS 2035 (see **Table 3**)
[3] Difference between initial (before) and final weight.
[4] Residual waxes extracted with chloroform from 20 P. *viburni* adult females after detergent spray.
[5] Means with different letters in a column are significantly different ($p \leq 0.05$) according to Tukey's test. Data extracted from Santibáñez [35].

Table 1. *Pseudococcus viburni* water losses and residual waxes after detergent sprays.

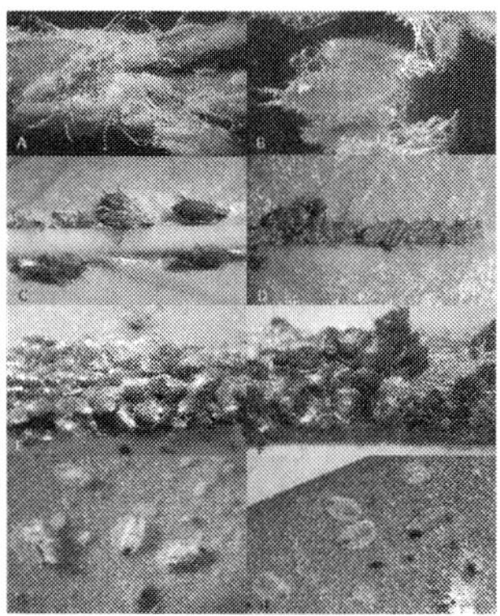

Figure 1. Healthy hemipterans before (left column) and either minutes or a few hours after exposure in 1–2% detergent solutions (right), presenting symptoms of dehydration, browning, body collapse, and wax removal. A, *Pseudococcus longispinus* Targioni and Tozzetti (Pseudococcidae); B, *P. longispinus* after 5-s immersion in SU 120 (see details in **Table 3**); C, *Aphis gossypii* Glover (Aphididae); D, effect of TS 2035 on *A. gossypii*; E, *Eriosoma lanigerum* (Hausmann) (Eriosomatidae); F, effect of TS 2035 on *E. lanigerum*; G, *Siphoninus phillyreae* (Haliday) (Aleyrodidae); H, *S. phillyreae* a few days after sprayed.

Detergents[1] (%, v/v)	% dislodgment[2] (D)	% mortality (M)	CD = 100×D/[D + M])
1.00%	22.2	59.4	27.2
0.50%	21.7	32.6	40.0
0.25%	14.1	17.6%	44.5
Control[3]	3.2	18.8%	14.5

[1] Quix solutions (see **Table 3**).
[2] Individuals found after immersion for 5 s + filtration.
[3] Tap water. Data extracted from Curkovic and Araya [37].

Table 2. *Panonychus citri* dislodgement (D), mortality (M) at 24 h, and contribution of dislodgement (CD) to the control (D + M), after immersion of infested lemon leaves in the laboratory.

1.2.2. *Arthropod dislodgement*

Detergents and soaps contain surfactants, that is, compounds that reduce the surface tension of solutions, enhancing their capability to wet and wash arthropods off. Thus, sprays can dislodge motile forms of phytophagous pests, as nymphs and adults of mites, thrips, etc. (particularly when the solution runoffs on the leaves). Even not necessarily all removed individuals die, and dislodgement causes significant reductions of populations infesting the foliage. Dislodgement has been highlighted as an anti-herbivore trait [36] that reduces their phytophagous performance on the plant. In a laboratory study, up to 22% dislodgment of the citrus red mite *Panonychus citri* McGregor (Acari:Tetranychidae) infesting lemon (*Citrus* × *limon* (L.) Burm.f.) leaves occurred after immersion in a detergent solution at 1% (v/v), significantly greater than water alone [37]. Mite mortality was also greater along with detergent concentration, but the relative contribution of dislodgment to total control (dislodgment + mortality) was even greater (44.5%) when the lower concentration (0.25%, v/v) was used (**Table 2**). In another study, 22% of the Chilean false red mite *Brevipalpus chilensis* Baker (Acari: Tenuipalpidae) were washed off vine leaves after immersion in a detergent (see **Table 3** for details) solution, but lower concentrations contributed less to total control [38], suggesting that dislodgement depends on the type of detergent and/or the mite species.

Not many reports have demonstrated dislodgment when soaps and detergents are used for pest control, although surfactants have been mentioned as useful tools to wash out arthropods plant substrates (including plant organs) for cleaning produce or pest sampling purposes [39]. For instance, ca. 28% of the western flower thrips, *Frankliniella occidentalis* (Pergande), were removed after the immersion in a 0.1% surfactant solution (see **Table 3**) from infested Coleus shoots (*Lamiaceae*), but the thrips were apparently not harmed [40].

1.2.3. *Drowning*

Arthropod respiratory system is formed by a net of conducts (traqueae) that allow direct gas exchange with tissues. It is connected to the exterior by spiracles that regulate opening by muscles [33]. The surfactant properties of detergents and soaps allow the solutions to enter the spiracles [41, 42]. The solutions fill the traqueae, causing drowning and death. No reports have been found describing this mechanism for pest control, but several papers have mentioned drowning as a mortality factor after surfactant sprays on insects and mites [43, 44]. In larger insects, this seems to be a lethal mechanism after exposure to D + S [43].

1.2.4. Other mechanisms

Interference with cellular metabolism [41], repellency [30], breakdown of cell membranes [42], abnormal juvenile development [12], caustic activity, uncoupling oxidative phosphorylation, and/or even nervous system disruption [45] have been also indicated as possible modes of action of D + S, but further details have not been found. Interestingly, in nature, surfactants have been highlighted as a mechanism of defense developed by some insects against their predators by producing oral secretions containing surfactants that, for instance, stop ants attacking beet armyworm, *Spodoptera exigua* (Hübner) caterpillars (Lepidoptera: Noctuidae). After exposure, the ants covered by the secretion are engaged in intensive grooming that persisted for a few minutes, enough to save the caterpillar. Besides, after cleaning, ants were reluctant to attack a second time [46]. In fact, the author has regularly poured pure dishwashing detergents (~5 mL on their path) to successfully stop ant columns at home.

1.3. Detergents and soaps used for pest control in agriculture

1.3.1. Formulations

Table 3 presents the characteristics and origin of 16 detergents and soaps used for pest control, or as co-adjuvants, reported in here. Many are liquids that perform better as insecticides and miticides [47], and a few are bars or powders. All were mixed in water to be applied, but bars needed, additionally, chipping and boiling before dilution. Several of the main world producers of cleaning products are represented in the list. About 44% of the products listed in **Table 3** correspond to either dishwashing, housecleaning, or personal cleaning products tested or used as alternatives to conventional pesticides. Thus, most products were not registered for pest control or agriculture use, but the results from research led later, in some cases, to the development of agriculture detergents (e.g., TS 2035 or SU 120 in Chile). Some D + S are produced locally, by relatively small producers, with raw materials easy to obtain, making suppliers and growers, particularly in developing countries, more independent from foreign surfactant producers. Information on D + S formulae was not always readily available and their components were not completely described, indicating only generically the type of compound (no chemical names given) or giving the range of the total surfactant content, but not precise figures. In fact, in many scientific publications reporting on the topic, there are no details on the specific inert ingredients or the surfactants (considered the active ingredients), or their respective proportions [47, 48].

Commercial names and formulations[2]	Companies[3] and countries	Surfactants (a.i.) and %[4] in c.p.	Declared uses[5] and references
Acco Highway Plant Spray Soap, L	Acme Chemical Company, PA, USA	Coconut oil soap[6] (38.5)	ASo, Moore et al. [63]
Break, L	BASF, Chile	Trisiloxane[7] (75)	Co, Sazo et al. [54]
Disolkyn, L	Bramell Ltda., Chile	Sodium disoctyl sulfosuccinate[6] (70)	Su, Sazo et al. [66]
Ivory Clear detergent, L	Proctor and Gamble, OH, USA	Acids salts of coconut oil and tallow[6]	HCD, Sclar et al. [69]
Key soap, B	Unilever, Ghana	Not provided	PCSo, Asiedu et al. [48]
LK dishwashing, P	Biotec S.A., Chile	Not provided	DiD, Arias et al. [47]
M-Pede, L	Mycogen Corp., CA, USA	Potassium salts of fatty acids[6] (49)	ASo, Butler et al. [30]
Nobla, P	Johnson and Diversey, Chile	Sodium alkyl benzene-sulfonate[6] (5-15)	HCD, Curkovic et al. [57]
Palmolive, L	Colgate-Palmolive S.A., Chile	Total fatty acids[6] (71)	PCSo, Arias et al. [47]
Quix, L	Lever S.A., Chile	Sodium benzene-sulfonate[6] (15-30%)	HCD, Curkovic et al. [34]
Safer, L	Agro-Chem, CA, USA	Potassium salts of fatty acids[6] (50)	ASo, Osborne and Petit [65]
SU 120, L	Johnson and Diversey, Chile	Sulfonates (14.9); lauryleter sulpfnate[6] (17.8)	AD, Ripa et al. [55]
Sunlight Dishwashing Detergent, L	Unilever, Ghana	LAS[6] (10-20) + sodium lauryl ether sulfate[6] (5-10)	DiD, Asiedu et al. [48]
Tecsa fruta, L	Protecsa, Chile	Xylene sulfonate[6] + nonylphenol[7] (1.5-2)	AD, Curkovic et al. [38]
Triton X, L	Sigma, MO, USA	Octyl-phenol hydrophobe series Polyethylene glycol ether[7]	ASu, Warnock and Loughner [40]
TS 2035, L	Pace Intl., Chile	15-17% sodium dodecyl sulfate[6], 4-6 ethoxilated alcohol%[7]	AD, Curkovic et al. [9]

[1] Not an exhaustive web search, thus, the characteristics were not found for all products; some of them can have different commercial names elsewhere.
[2] Liquid (L), powder (P) or bars (B).
[3] Fabricant or distributor at the time the original paper was published or current owner of the product.
[4] Either % w/v or v/v of surfactant(s) (considered the active ingredient) reported in the commercial product (c.p.) when available
[5] Reported use, housecleaning (HC), personal cleaning (PC), agriculture (A), horticulture (H), detergent (D), soap (So), or surfactant (Su) used as co-adjuvant (Co) or dishwashing (Di) detergent; bibliographical references where the product was cited
[6] Anionic surfactant.
[7] Non-ionic surfactant.

Table 3. Characteristic[1] and origin of some of the detergents and soaps reported herein.

1.3.2. Surfactants

The first synthesized surfactants were soaps, molecules with a relatively long hydrocarbon hydrophobic chain in one extreme, capable of binding lipids, and a hydrophilic carboxylic group in the other extreme bonded to either sodium or potassium [49]. Soaps are relatively easy to produce from natural raw materials (animal fat or vegetable oils). They were used in pest control as far back as the eighteenth century [50]. However, soaps did not perform efficiently in hard water (where they precipitate) or at low temperatures. Therefore, and also considering the shortage of raw materials in Europe after World War I, detergents were developed in the 1930s, overcoming the limitations of soaps [20], mainly by substituting the carboxylic end by a sodium sulfate or sulfonate, or other hydrophilic group. The main uses of both types of compounds worldwide are housecleaning (laundry and dishwashing), personal care (body washers, shampoos), but also in agriculture, food processing, etc. Today, the main raw materials used to produce surfactants are petroleum-based materials and plant oils (mainly from soybean and palm). The latter has an increasing production due to, among other factors, its low cost and toxicity, and natural origin. In fact, from the point of view of their use in agriculture, detergents, unlike soaps, cannot be used in organic farms because they are synthetic, nonnatural products. The recent changes in surfactant markets (including the need for safer, environmentally friendly, and economical products) have stimulated the production of new compounds. For instance, food and pharmaceutical processing surfactants or edible surfactants are available, providing alternatives that need to be tested as pesticides, besides older compounds [29, 51]. Surfactants in D + S reported herein are described in **Table 3**. In solution, surfactants tend to adsorb to the surface or interphase of materials, reducing hydrogen bridges between water molecules, thus improving their wetting capabilities. Besides, in contact with water, surfactants form micelles or small spheres, usually having the hydrophobic end inside, binding lipids, and the hydrophilic end outside. In this way, lipids are removed (degreasing effect) from the substrate and get diluted (solubilized). The electric charge of the hydrophilic end in solution can be neutral (non-ionic surfactants), negative (anionic, the most common among the D + S reported herein), positive (cationic), or both (negative and positive) [49]. Ionic surfactants can modify the pH of the solution. For instance, anionic surfactants tend to slightly acidify the pH, but they perform better at basic pH; therefore, the detergent formulae include some buffer agents. In fact, it was found that agriculture detergents (including all co-adjuvants) tend to alkalinize the solution in distilled water (pH: 7.8–8.9, depending on the concentration) [35], but only when above 1% (v/v) was prepared, maintaining the pH neutral otherwise [52]. In many cases, the surfactants vary between D + S formulations (in their chemistry and/or proportions), affecting their insecticide/miticide performance [38, 53]. Therefore, the activity of D + S needs some standardizing procedure in order to compare their activities as pesticides, for instance, comparing the proportion of surfactants (see below the case of some mealybugs), although differences can also be due to the particular type of surfactant, so this issue needs further research. Besides the house or personnel cleaning products, and some agriculture detergents, other sources for pest control are the co-adjuvants commercialized for specific functions, for example, wetting agents when mixed with pesticides or fertilizers in agriculture. Some of them have been individually or in mixtures tested as insecticides and miticides [52, 54].

1.3.3. *Efficacy as insecticides or miticides*

Most reports of pest control with D + S state relatively high levels of control (measured as either density reduction or mortality) against target pests. Those levels were usually achieved with the highest concentration tested, in most cases under or equal to 2%, either w/v or v/v, and considering the largest number of sprays [31]. The efficacy was directly related to coverage (the volume of water/ha used) and the stage of the pest (younger instars, except eggs, are the more susceptible ones, see **Table 4**) [11, 55]. In a few reports, however, the level of control obtained with soaps was poor [31, 56] or not significant when compared to some standard treatments (a recommended conventional pesticide). Maximum control was frequently measured when evaluations were conducted about a week after application, presumably due to a slower activity on arthropods than conventional pesticides [9], but some rapid stop-feeding response was also reported for insecticidal soaps, although mortality was achieved more slowly [12]. A few formulations include insecticides (e.g., pyrethrins are added in small amounts, [12]) for uses as agriculture soaps or louse shampoos [45], increasing their biocidal activity because of the addition of the natural neurotoxicant, but this is not the case of the products reported herein.

Detergents	LC_{50} on	LC_{90}[3] on
Tecsa fruta	1.4 b[2] (nymphs)	4.2 (nymphs)
	2.5 a (adults)	9.7 (adults)
SU 120	1.2 c (nymphs)	7.5 (nymphs)
	1.4 b (adults)	n/d[4] (adults)

[1] LC_{50} obtained by Probit analysis of data from commercial products in solutions (%, v/v) applied with a Potter tower (SU 120) or immersed 3 s in a solution (Tecsa fruta), values at 24 h after exposure.
[2] Means with different letters are significantly different based on Curkovic et al. [68].
[3] LC_{90}, values calculated from unpublished data, LC_{90} were 3–6× greater than the LC_{50}.
[4] No data are provided because maximum observed mortality was <50%.

Table 4. LC_{50}[1] and LC_{90} for *Myzus persicae* nymphs and adult females exposed to two agriculture detergent solutions.

1.4. Challenges and opportunities of detergents and soaps for pest control

1.4.1. *Phytotoxicity*

Toxicity to plants is a risk associated to the use of D + S, particularly at concentrations above 1–2% (v/v), but this effect should be a function of the proportion and type of surfactant(s) in the commercial formulation. It also depends on the plant species (its specific susceptibility or tolerance), their physiological condition, morphology, and growth stage. Phytotoxicity affects mainly leaves, flowers, and fruits [27, 57]; symptoms on the foliage range from yellowing to bronzing, and wilting or curling, up to necrosis and defoliation, whereas in fruits they range from small brown spots or massive epidermal browning to fruit dropping (**Figure 2**). Petal flowers can become brown or even necrotic when D + S are applied during flower bud appearance and blooming. These symptoms are also observed after repeated sprays with high concentrations (usually above 1%) of detergents [58] or when plants are under some type of stress (e.g., shaded plants, see below the case of *E. lanigerum*). It is believed that epicuticle wax removal in plants, at least in part, is responsible for this type of damage [34, 35]. Plant external cuticle is mainly made of cutin (one of two waxy polymers of long-chain fatty acids that are the main components of the plant cuticle, which covers all aerial surfaces of plants), and waxes that offer strong resistance to evaporation from the underlying cells [59]. These compounds can be removed by D + S, depending on the type and concentration [60], significantly increasing evaporation. Water losses can also occur through the stomata that have an opening regulated by guard cells [59] that are affected by some soaps, getting through their membranes [42]. Phytotoxicity has been observed more frequently in plants with pubescent surfaces (leaves), where the droplets act as lens causing burning [12], and lesser in those with heavily waxed leaves, limiting the use of D + S depending on the plant species and leaf anatomy. Regarding the pH of the sprayed solution, we have presented examples of data indicating only small changes not expected to be hazardous for plants when 1% or lower concentrations are used. In **Figure 2** (left picture), a recently set olive fruit (cv. Sevillana) presents browning on the lower half after exposure to an agricultural detergent (**Table 3**, [27]), even at 0.5% c.p. (v/v). The fruit later aborted, and the same happened in several other table (Kalamata, Manzanilla) and olive oil varieties (Arbequina). In fact, because of phytotoxicity, detergents (and horticultural oils) should not be applied to olive trees from flower bud to stone hardening (about 1-cm fruit diameter) [27]. Similar results have been observed in grapes at blossom and fruit set. These examples demonstrate specific susceptibility to surfactant sprays, since D + S can be applied at the same or even greater concentrations, on other fruit species during fruit set, without phytotoxic effects (e.g., apples and citrus). However, **Figure 2** (right picture) presents apple leaves damaged by weekly detergent sprays (n = 4, at 0.5% (v/v), see **Table 5**), probably due to the abnormal susceptibility of plants maintained for a long time (above a year) at a greenhouse covered by a shade mesh, before the trial was conducted. This condition might reduce the thickness of the cuticle layer and make the plant more susceptible to damage (sun burnt) or water losses [61]. In fact, foliage of apple trees in orchards sprayed with the same detergents (at 1%, v/v) did not present phytotoxic symptoms at all. Therefore, it is necessary to test at a small scale detergent and soap sprays, case by case, before being sure to conduct a larger-scale application. To do this, the evaluation should be conducted within a week or less after the spray for symptoms to be observed [27, 31], thus selecting tolerant species or adequate plant growth stages. On the other hand, phytotoxicity caused by D + S can be considered useful in crop protection, since some can be used directly as either herbicides or herbicide co-adjuvants [42].

Figure 2. Symptoms of phytotoxicity on recently set olives during the spring (October, left), and on apple foliage in the middle of the summer (February, right), after a spray with detergents (0.5% or 1%, v/v, respectively).

Treatments	# sprays Dafs[3]	% mortality
TS 2035 0.5%	1 (0)	15.2% e[4]
"	2 (0 and 7)	38.0% de
"	3 0, 7, and 14	62.8% bcd
TS 2035 1.0%	1 (0)	61.1% cd
"	2 (0 and 7)	84.4% abc
"	3 (0, 7, and 14)	90.5% ab
Chlorpyrifos[1]	1 (0)	94.4% a
Control[2]	3 (0, 7, and 14)	0.0% f

[1] Lorsban 75 WG was applied once on February 12, 2014 (= day 0), at 80 g c.p./hL.
[2] Tap water was applied every time.
[3] Total number of sprays during the 2-week period.
[4] Days after first spray (dafs) the successive applications were conducted.
[5] Means with different letters are significantly different (p≤0.05) according to Tukey's test.

Table 5. Mortality of *Eriosoma lanigerum* adults and nymphs infesting potted apple trees, with up to 3 weekly sprays of an agriculture detergent (at two concentrations) versus one spray of chlorpyrifos.

1.4.2. Lack of residual activity

Some reports state that insecticidal soaps are not persistent since they suffer rapid degradation [12]. However, some other studies on detergents or surfactants have demonstrated that their residues persist on the substrate after application. Triton X and Tween 80 (see **Table 3** for details), two surfactants used as co-adjuvants, produced persistent residues, at least a week after the spray on tomato fruits or tobacco leaves, respectively [60, 62]. Despite that, D + S residues do not have residual activity in terms of protection over time [31], which occurs only in solution [45], thus they are considered strictly contact pesticides (spray or topic exposure), some affecting the pest quickly [12]. Some soaps have been incorporated into a diet causing a slight mortality in the laboratory [56], showing some ingestion activity, but only at high concentrations (5× the recommended field rate). There is, however, some "residual" activity shortly after the application of D + S, if the solution lasts as either droplets or a liquid layer on the foliage and contacts the arthropod [47]. There is also the possibility of re-hydration if, for instance, relative humidity increases enough and shortly (after the spray) during fog events, to re-dilute D + S residues. It has been proposed to conduct repeated and frequent sprays of D + S to counteract their lack of residual activity on recurrent pests (see **Tables 5** and **6** for successful examples), but some concerns have been mentioned about the potential buildup of surfactants in the soil [63], although specific studies have not been conducted, except for some co-adjuvants [64]. On the other hand, the lack of residual effect turns out to be an advantage, preventing mortality of beneficial arthropods released after residues, which are dry, making D + S compatible with biological control and IPM programs.

Treatments	# sprays[3]	Dafs[4]	% mortality[5]
TS 20351[1]	1	0	29.0 cde
	2	0 and 10	23.7 de
	3	0, 10, and 20	51.7 abc
	4	0, 10, 20 and 30	54.2 ab
	1	30	49.6 cd
Imidacloprid[2]	1	0	78.8 a
Control	0	0	12.0 e

[1] At 0.5% c.p. (v/v).
[2] Confidor 350 SC applied once on January 24, 2013 (day 0), at 60 cc c.p./hL.
[3] Total number of sprays/treatment.
[4] Days after first spray (dafs) the successive applications were done.
[5] Means with different letters are significantly different ($p \leq 0.05$) according to Tukey's test. Unpublished data.

Table 6. Mortality of *Parthenolecanium corni* nymphs infesting vines, with one to four sprays (every 10 days) of TS 2035 at 0.5% versus one spray of imidacloprid.

1.4.3. Legal restrictions and registration

Authorization is an obligatory requirement to legally utilize D + S as pesticides in agriculture. It implies the demonstration of no toxicological risks (including ecotoxicology) and agronomic efficiency, based on science, excluding compounds that do not comply. The process requires a large effort, and it is slow and expensive, making the agrochemical industry to proceed only when the economic return is attractive. There are a few cases of registered D + S as insecticides and/or miticides for agriculture, a few in the United States [30, 65]. In Chile, there has been one registration (Disolkyn, see **Table 3**) for a few years during the mid-2000s [66], but it was not renewed, so there are no legally available D + S for pest control currently in this country. Despite that, non-registered D + S have been used in Chile for pest control, suggesting that they do not cause problems. Their use with no sanctions has occurred because this is an issue not regulated specifically, since the products can be declared as used, for instance, as tree cleaners (an authorized use in some agricultural detergents), pest control being the real purpose [11]. However, growers subjected to the certification process do not use D + S. This causes a serious bottleneck for registration and development for these compounds as tools for pest management. Besides, the chemical and agrochemical industry have not made large efforts for detergent registration as pesticides, in part for a low market expectative in economic terms (low profit), and also due to the difficulty and elevated costs involved. For D + S, government agencies require the same requisite used for the registration of conventional pesticides, making even more difficult for the industry to spend efforts in a registration process for these types of compounds. However, as mentioned before, many surfactants, detergents, and soaps are safe for the environment and the users, and some are even food additives or edible surfactants, so there is room for pesticide development to identify and select those D + S with very low risks. Similar to the case of horticulture oils, pheromones, or biological pesticides [12, 13, 18], D + S should be developed as safe products, obviously excluding those questioned and dangerous [15, 17]. Therefore, authorization for D + S must be addressed by all the actors involved: government (registration agency, Departments of Health and of Agriculture), producers (the surfactants industry and agrochemical companies, suppliers, and distributors), the academic sector (researchers from the agronomic, chemistry, and toxicology areas), and even grower and consumer organizations (particularly those advocated to consumption of safer foods). Only by acting jointly, the analysis, selection, and development will lead to register and use D + S in pest management. Once available, these compounds will serve in IPM, but also to conventional or organic production schemes, and serve in many complex scenarios (e.g., used very close to harvest with no other management options).

1.4.4. Spray conditions

Since D + S work strictly by direct contact, application should maximize the exposure of the pest as much as possible. Spray equipment must be adapted, for instance, modifying nozzles orientation in order to apply from underneath the leaves or fruits, where mealybugs, spider mites, or whiteflies use to feed [9]. Air-blast or powered backpack sprayers have been preferred for D + S applications, since better coverage and smaller droplets are achieved [9, 27]. If possible, trees might be pruned before spraying surfactants in order to increase pest exposure and air circulation that will help in the dehydration of treated insects and mites [9]. Solutions should be applied considering whole coverage of infested organs, using high volumes of water/ha and high-pump pressure during the spray [8, 63]. Besides, sprays should be done early in the morning or late in the evening to increase the duration of the wet layer and extend their insecticide lifetime [31].

1.4.5. Pest biology and ecology

The habits, biology, and morphology of the pest should also be considered to maximize exposure by D + S sprays. For instance, nocturnal pests (armyworms (Lepidoptera: Noctuidae) or snails (Mollusca: Pulmonata, Helicidae)) should be sprayed at night for direct exposure. In fact, some noctuids have not been controlled efficiently by diurnal soap sprays in the field [56]. For the greenhouse whitefly *Trialeurodes vaporariorum* (Westwood), nocturnal sprays were also recommended, since evaporation is low and adults are less mobile, being more likely reached by the solution [47], but diurnal application is efficient against the sessile stages (older nymphs). In pests known as susceptible to D + S, however, some specific instars are less (or not) vulnerable (e.g., spider mite eggs are less susceptible than mobile forms). In fact, in one report only slight activity against overwintering eggs of the European red mite *P. ulmi* (Koch) was found [67], while significantly greater summer eggs LC_{50} (1.5–2.3×) than adult females of the two-spotted spider mite, *Tetranychus urticae* Koch (both Acari: Tetranychidae), were observed in another study [68]. In the case of whiteflies, eggs and pupae are less susceptible, whereas nymphs or adults are severely affected by detergent sprays [27]. Mealybugs (pseudococcids) are difficult to reach by either contact or systemic insecticides in the field when they colonize fruit cavities, woodcuts, or roots [9]. In general, therefore, it is necessary to find the vulnerability for each pest species to be controlled with D + S.

2. Review of agriculture pests controlled with detergents and soaps

2.1. Hemiptera

Most examples of pest species controlled with D + S belong to this insect Order. They are the main target group because of their (a) size, being small (most), therefore highly dependent on their protective wax layer; (b) exposure on plant tissues, many being relatively easy to reach and/or remove from the foliage by the spray; (c) type of cuticle, being either soft or thin, thus more susceptible to D + S; (d) damage, as most species cause it when reaching high populations, thus, a significant reduction (but maybe not eradication) is enough to secure satisfactory yields, as expected for surfactants; and (e) null development of resistant populations as with conventional insecticides, thus, management with multisite D + S helps to avoid or reverse the problem, etc. The following review presents the most important hemipteran groups controlled with these types of compounds.

2.1.1. Aleyrodidae

Whiteflies are plant-sucking pests, having many generations per crop cycle, which infest mainly the foliage (usually the underside of leaves) of vegetables, tree fruit orchards, and ornamentals. They affect plant growth and yield by sap sucking, transmission of some diseases during feeding, and release of honeydew on the foliage and fruits, allowing the colonization by sooty mold. This fungus reduces both photosynthetic capacity and the value of the produce (downgrading the price of fruits and vegetables). Honeydew also serves as food for attendant ants that disturb biological control agents. Whiteflies have externally a conspicuous white-dusting wax layer to protect them from dehydration, also serving to reduce insecticide exposure. Detergent and soap sprays have been widely used to target the underside of the leaves and control whiteflies, despite some limitations against these pests as the lack of both systemic activity and residual effect. To counteract these narrowing factors, sprays require to be frequent, to cover the whole population. Besides, as whiteflies have several generations lasting about a month per crop cycle, each one should receive sprays. Butler et al. [30] were one of the first researchers in testing 16 D + S (e.g., M-Pede, Palmolive, etc.; see details in **Table 3**) on the control of the sweet-potato whitefly, *Bemisia tabaci* (Gennadius), 48 h after spraying several vegetable and ornamental species under greenhouse conditions. The production of honeydew by nymphs was measured as an evidence of nymph survival. In fact, there was a significant and inverse regression between the number of honeydew droplets (trapped on sensitive paper placed below infested leaves) and D + S concentration. The authors also found above 85% mortality (against the control sprayed with water) with 13 D + S at 1% either v/v or w/v, even under heavy infestation. Besides, adult whiteflies were removed from the leaves by the sprays and some ended adhered to the lower foliage and died. This is an additional control effect when using these types of compounds and it probably explains the reduction in adult's captures in traps after the application. D + S have also been tested on the greenhouse whitefly, *T. vaporariorum* (Westwood), but with dissimilar results. For instance, D + S were sprayed on infested seedling tomatoes (less than 10-leaf stage), and yield, plant toxicity, and nymph reduction were measured. Slight but significant nymph reduction (compared to the control) was observed when M-Pede insecticidal soap was sprayed at 2% (v/v), not causing yield losses. Weekly applications were suggested to control *T. vaporariorum* in tomato greenhouses, without plant toxicity risk [69]. Several other detergents (e.g., Ivory Clear detergent) used at 2% significantly reduced nymph density, but they also caused damage on the plant and yield reduction. In another study evaluating 12 D + S, only one product (LK dishwashing at 4–5% c.p. v/v) provided a bit over 50% adult *T. vaporariorum* mortality 24 h after the spray on infested bean plants (nymph mortality was not evaluated). Solid and liquid soaps caused similar results (below 35% mortality), except for Palmolive, that reached ~42% mortality using a 4% solution, but the lethal effect was dependent on the presence of liquid residues on the foliage [47]. High levels of control of the ash whitefly, *Siphoninus phillyreae* (Haliday), in olive trees sprayed with agriculture detergents (TS 2035 and *Tecsa fruta*) at 1–2% v/v, were reported [58]. Best results were obtained when detergents were applied on younger nymph stages (particularly nymph I) infesting potted plants, easier to cover with the spray. In another study in a pomegranate orchard, it was found that *S. phillyreae* nymphs I–III were easier to control whereas eggs and pupae (nymph IV) were far more difficult to kill with the same concentrations; thus, adults can emerge after the spray, not being a good predictor of whiteflies control [9]. Detergents at 0.5% (v/v) and above used against *S. phillyreae* produced toxic effects in Chilean olive orchards when used between flower bud and recently set fruits, precluding its use between those phenology stages, but pomegranates, on the other hand, were highly tolerant to 1% detergent (see **Table 3**) solution, and did not suffer either fruit or foliage damage. Results from other reports [9, 27] confirm that whiteflies are good targets to be controlled by D + S.

2.1.2. Aphidoidea

Aphids (*Aphididae*) are also very important plant-sucking pests, having impacts similar to whiteflies. Aphids tend to congregate on the buds and leaves, and also have several generations per season, but they tend to infest the foliage, twigs, and flowers during the spring, late in the season (close to harvest). Some cause the leaves to curl, forming refuges, being harder to reach with contact insecticide sprays, although this is easier with D + S due to its surfactant properties. Their bodies have less conspicuous wax layers than whiteflies. Because of that, sprays with contact insecticide target directly the colonies. D + S have been widely used to control aphids, with similar considerations as in whiteflies. Puritch [70], one of the oldest reports in recent times, found that soaps and their respective fatty acids (at 0.5% v/v) were active against the balsam woolly aphid, *Adelges piceae* (Ratz.) (*Adelgidae*), in Canada. The soap was more effective than the corresponding fatty acid, but both had neither ovicidal nor residual effect. Moore et al. [63] published one of the first reports of housecleaning products (D + S) used against three Aphididae species (the green peach aphid, *Myzus persicae* Sulzer; the spirea aphid, *A. spiraecola* [=*citricola*] (Patch), and the black bean aphid, *A. fabae* Scopoli), infesting several species of ornamentals. They found that Ivory liquid dishwashing at 1–2% (v/v) sprayed until runoff notably reduced populations immediately after the application. Plant toxicity was observed, particularly on plants with pubescent epidermis. In another report [69], the activity of two detergents on *M. persicae* nymphs and adults was evaluated by spraying (SU 120) or immersion (Tecsa fruta), in the laboratory. The last detergent was significantly more active (having a smaller LC_{50}, considering the smaller amount of active ingredients, that is, surfactants in the formulation) regardless of the aphid instar. Nymphs were more susceptible (**Table 4**: LC_{50} for adults was 1.2–1.8× greater).

Woolly aphids (*Eriosomatidae*) are also sucking pests that debilitate the host plant, release honeydew, and cause cankers. *Eriosoma lanigerum* infests roots but also the axils of leaves, twigs, branches presenting cuts, and occasionally the fruits. They produce large amounts of wax filaments, forming a woolly layer that serves as refugee for adults and nymphs, and protect them from sprays. They have up to 11 generations/season, and control is necessary when populations increase, mainly starting at the end of the spring up to harvest, and require repeated applications. Some unpublished data from a factorial experiment conducted on potted-infested apple trees in Chile demonstrated that both factors, detergent concentration and the number of sprays (one and up to three were contrasted in a 2-week period), were significant on *E. lanigerum* mortality, although no significant interaction was found. When comparing with the standard, results suggest the double spray of the 1% detergent solution (TS 2035, see **Table 3**) was as efficient as one application of chlorpyrifos, a residual insecticide (**Table 5**). Besides, three sprays at 0.5% achieved similar results as one or two sprays of the 1% solution, and these concentrations are alternatives if the greatest concentration causes plant toxicity. In fact, apple leaves were damaged by the treatments, probably due to the shade conditions in the greenhouse where the plants were grown, as mentioned before. Overall, woolly aphids were well controlled by the detergent.

2.1.3. *Coccidae*

Coccids or "soft scales" are important plant-sucking pests that infest mainly leaves and branches, and occasionally fruits, affecting plants similarly than whiteflies and aphids. Scales are relatively exposed to sprays, but their bodies are protected by a thick and hard shield. Because of that, sprays with contact insecticide target mainly young nymphs that have a poorly developed shield. Since coccids have usually one or two generations/year, the timing for insecticide contact sprays must be precisely defined by monitoring. Detergents and soaps have been informed to control coccid pests since several decades ago (e.g., Singh and Rao, 1979, on the green scale, *Coccus viridis* (Green) [71] in India), despite their lack of both systemic activity and residual effect. Reimer and Beardsley [72] found that an insecticide soap spray at 0.8% v/v caused significant reduction of *C. viridis* (~50% less scale survival compared to the control) infesting coffee trees (*Coffea arabica* L.) in the United States, 4 weeks after the spray. However, populations were significantly greater than those achieved with the standard treatment (fluvalinate had almost no scales at the same time), although no details on the scale population composition were provided. More detailed reports informed over 87% mortality (death individuals over the total) on the nymphal stages I and II of the black olive scale, *Saissetia oleae* (Olivier) infesting *Citrus × paradisi* (Macfad.) (grapefruit) and *Nerium oleander* L. (oleander), after the immersion of foliage recently colonized in two different housecleaning detergent solutions at 0.5–1% (either w/v or v/v, depending on formulations; Quix or Nobla, see **Table 3**) [34, 57]. However, similar mortality on adult females was obtained only with greater concentrations that caused plant toxic effects; defoliation and leaf necrosis was observed when above 2% was used [34]. In Brazil, nymphs and adult females of the pyriform scale, *Protopulvinaria pyriformis* (Cockerell), have also been controlled (over 77% mortality) by spraying a neutral detergent at 2% v/v, without causing toxicity on the dwarf umbrella tree, *Schefflera arboricola* (Hayata) Merr. [73]. More recently, satisfactory control of *S. oleae* young nymphs on commercial olive orchards sprayed with agriculture detergents at 0.5–1% v/v has been reported, but avoiding sprays during plant stages susceptible to toxicity (see above) [27]. The repeated use of detergents (two or more consecutive applications) has achieved a reduction in sooty mold and honeydew production (reducing the presence of attendant ants, thus improving conditions for biological control agents), and improved control of recurrent soft scales, somehow replacing detergents lack of residual effect [27]. For instance, a field trial conducted during the summer on a Chilean vineyard heavily infested with the European fruit Lecanium scale, *Parthenolecanium corni* Bouché, mainly targeting the first nymph instar of the second generation, obtained up to 54% mortality after up to four sprays (applied every 10 days) (details in **Table 6**). Results indicate that mortality increased significantly over time along with the number of applications, and no toxic effects on plants were detected even after four successive sprays at this concentration (0.5%). Interestingly, a single spray of detergent applied at day 30 provided 49% mortality (significantly similar to the three-spray treatment), but allowed the nymph population to develop and cause damage for over a month. These results suggest that a similar program of sprays, but at 1% (considered still safe for vines), might significantly improve control of this scale in vineyards. Besides, this program of repeated applications would also serve to control important and synchronic pest as aphids, mealybugs, thrips, and spider mites, all susceptible to soaps and detergents.

It is worth noting that other coccid species have been reported to be satisfactorily controlled by D + S: the soft scale *Ceroplastes* spp., Sabine, 1969 in Australia [74], the oldest report on the use of surfactants alone as insecticide during recent times; the pine tortoise scale, *Toumeyella parvicornis* (Cockerell), in the United States [75], whereas a few scales species have not been controlled, for example, a spray of an insecticidal soap on the calico scale, *Eulecanium cerasorum* (Cockerell), an invasive pest of shade trees in the United States, was rated as relatively ineffective [76]. Overall, these results indicate that coccids are good targets for D + S, but the responses vary between species and that they depend on the management strategy.

2.1.4. *Diaspididae*

Armored scales are also sucking pests, but have a dorsal and protective shield not glued to the body. They colonize mainly branches and fruits (and eventually the leaves) of tree fruit orchards and ornamentals, but do not produce honeydew. Only nymphs I are mobile (crawlers), but once they set on a structure, they lose their legs and become sessile. Diaspidids have usually two to four generations/year. Reports on armored scale control with D + S are less frequent. For instance, the mortality of the oleander scale *Aspidiotus nerii* Bouché nymphs recently set in the wood, 1 week after application on a Chilean olive orchard, reached 60–70% with two agriculture detergents (an horticulture oil reached less than 40%), whereas chlorpyrifos provided close to 100% mortality ([58], **Figure 3**). However, the level of mortality with the detergent spray declined dramatically the next weeks because of the lack of residual effect, whereas mortality with chlorpyrifos kept almost unchanged 5 weeks later due to its long residual effect. The control of the white mango scale, *Aulacaspis tubercularis* Newstead, with housecleaning detergents was evaluated on mangoes (*Mangifera Indica* L.) in Mexico, achieving significantly less colonies and scales on the leaves in comparison with the control, 1 week after the spray. However, the population density increased rapidly 2–3 weeks after application (due to the lack of residual effect), being greater than those obtained with conventional insecticides [77]. The effect of six housecleaning soaps (details in reference) sprayed on the cycad aulacaspis scale, *A. yasumatsui* Takagi, infesting cica crops (Cycadaceae) in Costa Rica, caused significantly smaller densities of crawlers and significantly greater numbers of dead females after the application of soaps (~3%, v/v) with a backpack sprayer [78]. However, the spray of some detergents significantly reduced the activity of entomopathogenic fungi (*Metarhizium*) as well. These results also suggest that surfactants might act as contact fungicides against some plant fungal disease agents.

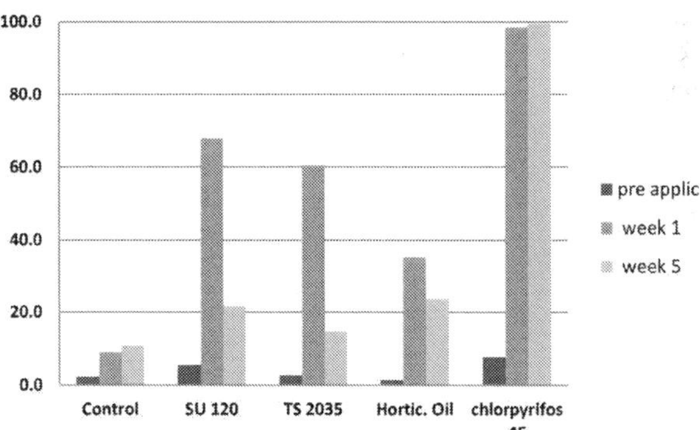

Figure 3. Mortality (%) of *Aspidiotus nerii* nymphs I (about 1000 individuals were counted per treatment/date) infesting olives sprayed with 10 L of detergent solution/tree, with two products (SU 120 and TS 2035, both at 1%, v/v), a horticulture oil (1%, v/v), and chlorpyrifos (at the recommended rate: 120 mL c.p./hL). Evaluations were conducted right before (pre-application: natural mortality was <10%) and 1 and 5 weeks after application, Copiapó, Chile. Data extracted from Curkovic and Ballesteros [58].

2.1.5. *Pseudococcidae*

Mealybugs are, in general, similar to soft scales regarding the effect on infested plants. However, mealybugs do not significantly reduce plant growth and tend to infest fruits and branches instead of leaves, wood crevices and cuts, zones of fruit contact and calyx cavities, where they can stay even after harvest. Some are serious quarantine problems for exports. Because of that, detergents or soaps are usually not used for mealybug control in orchards oriented to export (however, see the use as postharvest treatment below). Consequently, an intense chemical control program is applied in Chilean orchards exporting fresh fruit, using conventional insecticides (preferring systemic and/or residual products), but their efficacy is still relatively low. This is due mainly to the insect's habits (see Section "Pest biology and ecology"), its phenology (having three to four generations/season they infest the plant the entire season), and morphology (mealybugs are superficially covered and protected by a layer of waxes and woolly filaments). When exposed to sprays, however, mealybugs are highly susceptible to contact insecticides, including D + S. For instance, two agriculture detergents, Tecsa fruta and SU 120, were compared to control *P. longispinus* under laboratory conditions, finding significantly lower LC_{50} values (~1.9%, v/v) in the latter commercial product versus ~18% (v/v) in Tecsa fruta, which contains almost 8× less, and different surfactants. Besides, this study also showed that younger individuals (nymphs II) were significantly more susceptible than adult females and mortality was greater when more coverage (spray volume) was used. Interestingly, at greater concentrations and spraying volumes, mealybugs were glued to the surface by waxes removed from the cuticle and deposited under their bodies [8]. In a more recent study, a significant reduction (82.4% in a 2-year average from the control) in mealybug densities (the cotton mealybug, *Phenococcus solenopsis* Tinsley, and the papaya mealybug, *Paracoccus marginatus* Williams and Granara de Willink) per cotton plant (*Gossypium hirsutum* L.) after a 0.1% (w/v) powder detergent solution (no data on name or composition provided) was obtained after spraying eight times during two seasons in India. The detergent was overcome only by acephate and chlorpyrifos treatments, which reached above 95% reduction, but the detergent provided better control than several entomopathogens (*Beauveria, Metharizium*) and neem oil (ranging from 23 to 69% reduction), and was more selective to ladybirds (Coccinellidae) and spiders [79]. In another two reports, a significant control of the citrus mealybug, Planococcus citri (Risso) was found, by soaps [80], or by D + S on white yam (*Dioscorea cayennensis* subsp. *rotundata* (Poir.) J. Miége) 14 days after being sprayed with a detergent (sunlight at ~1.9% c.p.) and a soap (key soap at ~2.5% (w/v) c.p.). Mortality of *P. citri* on stored white yam was above 92% whereas soybean oil reached 99%, and cypermethrin and imidacloprid provided total control. The authors concluded that all detergent, soap, and soybean oil treatments are alternatives that need further research as postharvest treatments [48]. In a few reports, mealybug control with detergents and soaps has not been successful, in part attributable to the formulation tested, probably under ongoing efforts for development, but all articles emphasize the need to continue the research to identify conditions to obtain better control [81]. There are reports of species belonging to other hemipteran families controlled with D + S: *Capulinia* sp. (*Eriococcidae*) [82] in Venezuela, and the mobile scale of the olive tree, *Praelongorthezia olivicola* (Beingolea) (*Ortheziidae*) [83] in Chile, somehow confirming Hemiptera as the main target group for these types of products.

2.2. Thysanoptera

Thrips are serious pests of vegetables, flowers, and fruit orchards, mainly affecting cut flowers and the skin of fruits (causing russet). They can produce silvering on flowers, leaves, and fruits, downgrading their value. Adults and nymphs are not sessile but tend to stay inside flower structures, under sepals, or at the contact point between either fruits or leaves and fruits. Therefore, D + S can be useful resources to reach them at those protected sites, by being used alone or as co-adjuvants (as surfactants) for conventional insecticides. However, trials evaluating thrips control have achieved different results in terms of mortality or density reduction when D + S have been used alone. For instance, the use of an agricultural soap (Acco Highway plant spray soap at 1% v/v) on the greenhouse thrips, *Heliothrips haemorrhoidalis* (Bouche) (Thripidae), caused important reductions on populations on infested ornamental plants (*Acacia longifolia* (Andrews) Willd; *Pittosporum tobira* (Thunb.) W.T. Aiton) after every spray ($n = 4$ in a 3-week period), but not after only one spray [63]. On the other hand, almost negligible mortality on the Western flower thrips, *Frankliniella occidentalis* (Pergande) (Thripidae) has been reported after exposed to a soap solution (Soapline containing 60% potasic soap, Syngenta Agro, Spain) by either a residual (treated leaves were sprayed before exposure, so no control should be expected) or a topical bioassay [84]. In the latter case, results were assumed to be due to a low efficacy of the soap used, but it might also be due to the short exposure time. These results suggest again the possibility of differential responses to D + S in distinct species from the same insect family.

2.3. Acari

Spider mites feed mainly on the content of epidermal and parenchymal plant cells. While feeding, they do not reach vascular vessels; therefore, they do not produce honeydew. However, high populations can quickly develop on leaves causing bronzing, necrosis, and defoliation due to cell damage and the release of toxic substances. Mites tend to colonize the underside of leaves, where they need to be sprayed with contact and residual miticides, since colonization (for instance, from overwintering sites to the foliage developing during the spring) can last several weeks. During their development, they have sessile phases (proto- and deuto-nymphs), otherwise they are considered mobile arachnids. Besides, spider mites have several generations a year, being necessary to repeatedly control them along the season when populations reach dangerous densities. Some of the first modern reports of D + S used to control agricultural pests are related to spider mites [63, 65].

2.3.1. Tetranychidae

Osborne and Petit [65] found that the lowest insecticidal soap concentration (Safer at 1.25%, v/v) was effective in controlling adults and eggs of *T. urticae*, but also killed adults of the predatory mite *Phytoseiulus persimilis* Athias-Henriot (Phytoseiidae), although not their eggs. Therefore, predators can be used in conjunction with applications of low concentrations of soaps, giving better control than either tactic alone, provided that the release of the biological control agents is conducted after the spray. More recently, a significant effect of the agricultural detergent Disolkyn (at 0.1 and 0.15%, v/v) on *P. ulmi* (including some ovicidal effect) was found on severely infesting apple (*Malus domestica* Borkh.) trees in Chile, with a lesser effect onto the predatory mite *Neoseiulus californicus* (McGregor) (Phytoseiidae) (*N. californicus* had a good population recovery after the spray) in comparison with the standard treatment having residual effect (Pyridaben) [66]. In another study, the detergents SU 120 and Tecsa fruta were evaluated on mortality of *T. urticae* eggs and adult females, set on double-sided tapes placed on a slide immersed in detergent solutions in the laboratory [68]. The former product was significantly more active killing mites (smaller LC_{50}) for both instars. Eggs were significantly less susceptible than adults to both detergents. In another study, TS 2035 and M-Pede (**Table 3**, [9]) were sprayed at 1% (v/v) on a population of *Oligonychus* sp. mobile forms severely infesting a pomegranate orchard (**Table 7**). Both surfactants were statistically as efficient as the standard treatment (Pyridaben). Evaluations 2 and 9 days after the spray showed that populations did not recover in D + S treatments, whereas they were significantly greater in the control, causing subsequent damage on the foliage and fruits. No significant plant toxicity was observed. A horticulture oil also performed satisfactorily, but it had a slightly greater recovery of the mite population by day 9 [9].

Treatments[1]	**Days after a spray**		
	0	2	9
M-Pede	60.0 a[3]	17.5 ab	20.8 b
TS 2035	54.8 a	11.5 ab	9.3 b
Horticulture oil	56.0 a	12.0 ab	33.8 ab
Pyridaben[2]	53.3 a	1.8 b	6.5 b
Control	127.3 a	138.3 a	199.3 a

[1] Surfactants and oil (Ultraspray) at 1% c.p. (v/v).
[2] Sanmite 20 WP applied at 75 g c.p./hL.
[3] Means with different letters within a column are significantly different ($p \leq 0.05$) according to Tukey's test. Extracted from Curkovic et al. [9].

Table 7. Densities of *Oligonychus* sp. mobile forms on pomegranate before (day 0 = April 11, 2013) and after a spray (days 2 and 9) with several miticides.

2.3.2. *Tenuipalpidae*

A recent report indicates that the detergent SU 120 at 1.5% (v/v) sprayed in an infested vineyard had a significant effect on reducing *B. chilensis* mobile stages, particularly during the summer. Density reduction was not significantly different from the standard miticide acrinathrin. Mite recovery was observed almost 1 month after the spray, but eggs were apparently less affected [38].

2.4. Detergents and soaps used against other organisms

Other insects than those addressed herein, as armyworms (Lepidoptera: Noctuidae, [56]), cockroaches (Blattodea: Blatellidae [43]), and ants (Hymenoptera: Formicidae [44]) have been reported as controlled by D + S, or at least affected. Besides, the control of other organisms including mollusks [85] and fungi [86] with D + S or surfactants has also been reported. All this evidence demonstrates that the potential target for this type of control tactic is far beyond sessile, soft integument, and small insects or spider mites.

3. Costs and economic benefits of using D + S

Costs of detergents or soaps used against agricultural pests, in general, should be relatively low per spray (and it will become even lower if D + S increase their use in agriculture), but there are some exceptions (e.g., expensive insecticidal soaps sold in smaller containers for garden pests in the United States). **Table 8** compares the direct costs of applying a detergent program versus a conventional insecticide, considering having a residual effect shorter or similar to the period of evaluation in the field, and conditions where both strategies have achieved statistically similar levels of control for two pests in either apples or vines (see **Tables 5** and **6**). When comparing the detergent program versus chlorpyrifos used against the apple woolly aphid, the TS 2035 program cannot outcompete the conventional insecticide, being more than 2× more expensive. If Lorsban 4E be used (another much inexpensive chlorpyrifos formulation recommended at 120 mL/hL, with a cost of US$9.7/L), the cost of the detergent program would be about 3× more expensive. However, if other insecticides as buprofezin (Applaud 25 WP used at 120 g p.c./hL, US$42.1/kg) or imidacloprid (Confidor 350 SC) are used (modern and less restricted insecticides, but also more expensive products), considering application conditions and assumptions as described for chlorpyrifos, the standard strategy/detergent program ratio would increase, to near 0.79 (the detergent program being now only 20% more expensive than Applaud) and 1.49, respectively. In the latter case, the detergent program was 49% cheaper (including costs of products, equipment, and workers) than the conventional neonicotinoid. Now, when comparing the use of a neonicotinoid in vines against scales versus the detergent program, results also become very competitive in favor of the detergent strategy (ratio = 1.63). Even considering increasing the detergent concentration to 1% (see discussion in **Table 6**), the detergent program (three sprays) would be 1% less expensive than the use of imidacloprid once. Thus, detergents tend to be competitive when new, more expensive molecules, are used as standard treatments, a trend expected in the next years. The two main factors increasing costs of detergent treatments have been (1) the need to re-apply in order to counteract the lack of residual effect to achieve a level of control similar to that of conventional (and residual) pesticides. Thus, the cost rises due to the increasing value of motorized equipment and drivers, used two to three times (against just one application of the standard); (2) the use of concentrations about 8× greater than conventional pesticides to obtain similar results (detergents need to be used at 0.5–1% c.p. vs. the standards used at 0.06% (imidacloprid) or 0.12% (v/v chlorpyrifos or w/v buprofezin). Besides, since D + S must be applied using high volumes at relatively high concentrations, the amount of product used is larger. The examples presented are based on particular conditions (see the **Table 8** legend). However, the costs should vary among different countries, crops, management strategies, pesticide values, or pest species.

Pest species and crops	#[1] of detergent sprays ≈ to standard[2] control	A: US$ cost of standard (appl./ha)[3]	B: US$ cost of detergent (appl./ha)[4]	Ratio A/B[5]
Parthenolecanium corni on vines	3	113.5	23.2	1.63
Eriosoma lanigerum on apples	2	74.4	77.0	0.48

[1] Minimal number of detergent sprays necessary to achieve mortality not significantly different from the standard treatment (see details in **Tables 5** and **6**).
[2] Standard treatments; one application with imidacloprid (vines) or chlorpyrifos (apples) provided the best control during the period of evaluation.
[3] Cost of application + insecticide in Chile; considering 1 h of equipment (tractor + air-blast sprayer owned by the grower + the driver salary) to cover 1 ha (US$20 for apples or US$8.9 for vines, figures provided by growers); cost of insecticide product for either Confidor 350 SC used at 60 mL p.c./hL in vines (US$174.3/L), or Lorsban 75 WG used at 80 g p.c./hL in apples (US$34/kg), as standard treatments, prices provided by local suppliers.
[4] Cost of application of detergent TS 2035 (US$2.85/L), at 0.5% (v/v) for vines (coverage of 1000 L/ha), or at 1% (v/v) for apples (2000 L/ha).
[5] Ratio between the cost of the standard treatment/detergent program; when greater than 1, the detergent strategy is proportionally more convenient.

Table 8. Comparison of costs (US$) for detergent programs versus conventional insecticides, both as efficiently used to control *Parthenolecanium corni* in vineyards and *Eriosoma lanigerum* in apple trees.

It is important to point out that the exercise above does not consider other benefits of using D + S (used instead of conventional pesticides), as the avoidance of both pest resistance development to chemical pesticides or pest resurgence, or the relative improvement of the environment and the agro-ecosystem, or the reduction of risks of human intoxications (workers and consumers), and so on, because their costs are difficult to estimate. Therefore, if all those costs were valuable, it would probably make the figures much more favorable for D + S. Additionally, the access to markets preferring food not treated with conventional pesticides might also be considered an economic benefit. For instance, IPM or organic products can eventually achieve higher prices than conventional agriculture produce. Besides, foods treated with soaps or detergents will not have major restrictions to reach many different countries since they do not present questionable residues, making easier (and cheaper) the marketing process. In favor of conventional pesticides, an additional economic benefit of their use is their wider spectrum of action against some pest complexes in some crops, but D + S have also demonstrated an extended range of action on pests. Besides, some conventional products can protect for long periods against pests. However, some cannot be used during some phenological stages (Lorsban 4E is used today mainly as postharvest or winter treatment).

Among other examples in the literature, an IPM program was cost-effective at most of the studied sites where the majority of pest were controlled using spot sprays of insecticidal soap or horticultural oil versus the management with conventional pesticides applied on the whole plantation [87]. Another report showed that up to five detergent sprays could be applied before reaching the cost equivalent of controlling pests with conventional pesticides applied twice (only considering the value of the commercial product, but no other application costs) [11]. Similarly, a recommended mixture of a miticide plus the synergic surfactant co-adjuvant Silwet 77 was over 5× more expensive than the cost of using the surfactant alone, which provided most of the control. Unfortunately, the surfactant was not registered as miticide, and was not allowed as a legally authorized control method [53]. Reduced pest control costs, by the use of soaps, were also mentioned by Lee et al (2006) [88].

4. Detergents as insecticide co-adjuvants

The use of surfactants, including D + S, as adjuvant, improves both the active ingredient solubility in the formulation and its physical and biocidal performance (e.g., wetting properties on plant or insect cuticle). Co-adjuvants are added directly to the tank before applications with the same purposes [11]. The oldest report of using soaps (as co-adjuvant) in mixture with other pesticides in the tank was published in Australia in 1969 [74], as a part of the phytosanitary program in Citrus, providing a satisfactory degree of both, coccids and diaspidids control. Later, surfactants were described as co-adjuvants, particularly for cuticle penetration in insects [89]. Last year, an entomopathogen spore suspension (*Metarhizium anisopliae* strain M984) was tested, at the same concentration with or without the addition at the tank of an agricultural detergent (TS 2035; at a nonlethal concentration for *P. viburni* = 0.001%, v/v). A significantly increased mortality of *P. viburni* after the spray was obtained with the mixture (*M. anisopliae* + detergent), whereas the insecticide alone (not mixed with the detergent) provided significantly lower mortality (greater transformed LC_{50}, see **Table 9** [52]). Results show about one order of magnitude of differences in favor of the mixture of spore suspension with the detergent. These results justify the addition of detergents or surfactants during the formulation of commercial products, but they also open chances to reduce rates of pesticides used in the field when D + S are added to the solution in the tank. However, this hypothesis needs to be further tested.

Treatments[1]	**Time (h)**[3]	**LC_{50}**[4]
M. anisopliae + TS-2035[2]	24	8.8×10^6 ab[5]
M. anisopliae	24	8.6×10^7 c
M. anisopliae + TS-2035	72	7.8×10^6 a
M. anisopliae	72	3.3×10^7 c
M. anisopliae + TS-2035	144	6.1×10^6 a
M. anisopliae	144	3.0×10^7 bc

[1] Suspensions (2 mL) of *M. anisopliae* were sprayed/replicate ($n = 4$)/treatment (15–20 P. viburni adult females/replicate), using a Potter tower ST-4.
[2] TS 2035 at 0.001% (v/v).
[3] Three evaluation times were considered given the relatively slow activity reported for *M. anisopliae* on mealybugs.
[4] LC_{50}values were transformed to the respective amount of *M. anisopliae* CFU/mL.
[5] Means with different letters are significantly different ($p \leq 0.05$) according to Tukey's test. Extracted from Villar [52].

Table 9. *Pseudococcus viburni* LC_{50} values of a *Metarhizium anisopliae* strain M984, with or without the addition of TS-2035 at 0.001% (v/v) at different times after spray.

5. Postharvest control of pests with detergents

Immersion of the fruit in warm water has been used as postharvest pest control against several pests on diverse fruit species [90, 91]. Besides, several D + S are allowed for postharvest uses, including fruit cleaning. The combination of both approaches (warm detergent solution) was tested, finding that pomegranates infested with mealybugs and immersed in a 1% (v/v) TS 2035 solution (at 47°C) for 15 min, maintaining the pH at either 5.5 and 8.5, notably (but not totally) controlled *P. viburni*, a pest with quarantine status, usually found in the calyx cavity in postharvest (**Table 10** [92]). There were no adverse effects of the treatments on fruit quality; therefore, further evaluation of these factors at greater levels should be conducted to obtain total control, eventually becoming in an alternative to fumigation.

Temp.[1] (°C)	Det. Conc.[2]	pH[3]	Exposure time (min)[4]	Adult females	Nymphs II and III	Nymphs I	All mealybug stages
15 ± 2	0	5.5	15	2.75[5]	8.50	8.75	20.00
15 ± 2	0	8.5	15	2.00	6.25	22.00	30.25
15 ± 2	1	5.5	15	3.50	3.25	11.00	17.75
15 ± 2	1	8.5	15	2.25	7.75	18.00	28.00
47 ± 2	0	5.5	15	1.25	7.50	9.50	18.25
47 ± 2	0	8.5	15	6.00	4.75	15.25	26.00
47 ± 2	1	5.5	15	0.50	0.25	12.75	13.50
47 ± 2	1	8.5	15	1.00	1.25	2.75	5.00

[1] Water temperature.
[2] % TS 2035 c.p., v/v.
[3] pH corrected from neutral to acid (by adding phosphoric acid) or basic (by adding sodium hydroxide).
[4] Time pomegranates were immersed in solution (minutes).
[5] Means of selected treatments, showing greatest contrasts. Extracted from Carpio [92].

Table 10. Survivals of *Pseudococcus viburni* mobile stages (adult females, nymphs, and total), after postharvest immersion in detergent solutions, plus 1-month cold storage at 5°C, followed by 24 h at room temperature.

6. Conclusions and prospects

Many different agriculture pests (mainly hemipterans and spider mites) are efficiently controlled by detergents and soaps, provided they are directly covered by the spray. The knowledge of their biology and ecology must be used to improve their performance by increasing the pest exposure. The research on new potential targets and the combination of D + S with biological control agents should be studied. D + S can be used as well to avoid or even reverse pest resistance problems.

The modes of action of D + S as insecticides and/or miticides seem to be mainly wax removal, arthropod dislodgement, and drowning, but it is an unsolved issue in many situations yet. It is then necessary to keep researching on this issue to optimize the use of surfactants as pesticides.

Despite some environmental and toxicological concerns, the appropriate use of D + S, and the selection and formulation of surfactants with minimum risks (for instance, among the offer of new, safe, and low-cost surfactants), makes them potentially useful pesticides, but it is necessary to confirm their relatively safety (for mammals and the environment) and capacity for pest control, in food products.

There is a need to standardize the biocidal activity when comparing D + S, maybe based on the proportion of surfactants in the formulae or contrasting with some standard compound.

Detergents and soaps can be used as co-adjuvants (in the tank) for conventional or biological pesticides. D + S can also be applied first to debilitate pest insects and mites, spraying later insecticides and miticides. In both cases, a rate reduction for conventional (and more expensive and restricted products) is possible, but these issues need further research.

Detergents and soaps can be used in orchards, vegetables, or greenhouses, serving to conventional, IPM, or organic growers, making possible to reach highly selective markets and consumers willing to pay for foods free of insecticide residues and, at the same time, take advantage of their relative sustainable status, replacing conventional pesticides. D + S could be applied very close to harvest, when conventional pesticides cannot, due to the insufficient preharvest intervals.

However, in order to provide satisfactory control and become a greater tool for pest control, D + S need to solve the (a) lack of residual effect, (b) potential for plant toxicity, (c) legal status, and (d) cost. For multivoltine pests, or those infesting crops for long periods, their repeated use over relatively short periods has probed in several cases to provide a control equivalent to conventional (and residual) insecticides. Plant toxicity has been diminished by selecting tolerant crops, or tolerant phenology stages of the crops, excluding otherwise the use of D + S. This issue needs more research to identify tolerant crops and the conditions and mechanism causing plant toxicity, in order to develop safer D + S. The repeated applications of small concentrations of D + S have overcome these two problems, becoming useful tools for IPM productive schemes, particularly considering their multi-site action, selectivity to beneficial organisms, lack of residual effect, and relatively low environment and human toxicity. The facts that D + S are relatively quick to control, easy to produce and use, versatile, and lack major legal restrictions just improve their possibilities to be incorporated in pest programs.

The cost of efficient programs of control with D + S can be competitive with conventional pesticides, depending on the crop, pest, type of grower, and alternatives of pesticides, and it deserves a more detailed analysis, including the precise valorization of several benefits associated to the use of D + S, although some of them are difficult to measure, as lower probability of inducing insecticide resistance or pest resurgence, lower risks of intoxications to workers, etc.

Besides the cost issue, the authorization of D + S as pesticide products seems to be the next main challenge, being necessary that the industry (producers and suppliers), government agencies (regulatory apparatus), scientists (agronomists, entomologists, chemists, toxicologists), and even growers and consumers interact in order to develop a regulation process that allows to increase D + S registrations, particularly those safer compounds, that can be efficiently used with minimum risk (by far lower than conventional pesticides) at pre- and postharvest, becoming valuable tools for sustainable pest management.

Author details

Tomislav Curkovic S.

Address all correspondence to: tcurkovi@uchile.cl

Department of Crop Protection, University of Chile, Santiago, Chile

References

[1] Gill HK, Garg H. Pesticide: environmental impacts and management strategies. In: Solenski S, Larramenday ML, editors. Pesticides-Toxic Effects. Rijeka, Croatia: Intech; 2014. pp. 187–230.

[2] Kogan M. Integrated pest management: historical perspectives and contemporary developments. Annual Review of Entomology. 1998;43:243–270.

[3] Flint ML. IPM in practice: principles and methods of integrated pest management. 2nd ed. California: UCANR Publications; 2012. 293 p.

[4] Fagan LL, McLachlan A, Till CM, Walker MK. Synergy between chemical and biological control in the IPM of currant-lettuce aphid (*Nasonovia ribisnigri*) in Canterbury, New Zealand. Bulletin of Entomological Research. 2010;100:217–223.

[5] Echegaray ER, Cloyd RA. Effects of reduced-risk pesticides and plant growth regulators on rove beetle (Coleoptera: Staphylinidae) adults. Journal of Economic Entomology. 2012;105:2097–2106.

[6] Trumble JT, Alvarado-Rodríguez B. Development and economic evaluation of an IPM program for fresh market tomato production in Mexico. Agriculture, Ecosystems and Environment. 1993;43:267–284.

[7] Reddy GVP, Guerrero A. New pheromones and insect control strategies. Vitamins and Hormones. 2010;83:493–519.

[8] Curkovic T, Burett G, Araya JE. Evaluation of the insecticide activity of two agricultural detergents against the long-tailed mealybug, *Pseudococcus longispinus* (Hemiptera: Pseudococcidae), in the laboratory. Agricultura Técnica. 2007;67:422–430.

[9] Curkovic T, Ballesteros C, Carpio C. Integrated pest management in pomegranate [in Spanish]. In: Henríquez JL, Franck N, editors. Basis for the cultivation of Pomegranates in Chile. Santiago: Volumen 25 Serie Ciencias Agronómicas Universidad de Chile; 2015. pp. 159–232.

[10] Ehler LE. Integrated pest management (IPM): definition, historical development and implementation, and the other IPM. Pest Management Science. 2006;62:787–789.

[11] Curkovic T. Advances in pest control with detergents. A tool for integrated management [in Spanish]. Aconex. 2007;94:11–17.

[12] Weinzierl RA, Henn T. Botanical insecticides and insecticidal soaps. In: Leslie A.R., editor. Handbook of Integrated Pest Management for Turf and Ornamentals. USA: Lewis Publishers; 1994. pp. 538–590.

[13] Schwuger MJ, editor. Detergents in the Environment. New York: Marcel Dekker, Inc.; 1996. 336 p.

[14] Travis MJ, Wiel-Shafran A, Weisbrod N, Adar E, Gross A. Grey water reuse for irrigation: effect on soil properties. Science of the Total Environment. 2010;408:2501–2508.

[15] Kuhnt G. Behavior and fate of surfactants in soil. Environmental Toxicology and Chemistry. 1993;12:1813–1820.

[16] Jensen J. Fate and effects of linear alkylbenzene sulphonates (LAS) in the terrestrial environment. Science of the Total Environment. 1999;226:93–111.

[17] Soares A, Guieysse B, Jefferson B, Cartmell E, Lester JN. Nonylphenol in the environment: a critical review on occurrence, fate, toxicity and treatment in wastewaters. Environment International. 2008;34:1033–1049.

[18] Pavela R. Possibilities of botanical insecticide exploitation in plan protection. Pest Technology. 2007;1:47–52.

[19] Conti E. Acute toxicity of three detergents and two insecticides in the lugworm, *Arenicola marina* (L.): a histological and a scanning electron microscopic study. Aquatic Toxicology. 1987;10:325–334.

[20] Effendy I, Maibach H. Detergents. In: Chew A, Maibach H, editors. Irritant Dermatitis. Germany: Springer; 2006. pp. 249–256.

[21] Mapp CE, Pozzato V, Pavoni V, Gritti G. Severe asthma and ARDS triggered by acute short-term exposure to commonly used cleaning detergents. European Respiratory Journal. 2000;16:570–572.

[22] Pedersen LK, Held E, Johansen JD, Agner T. Less skin irritation from alcohol-based disinfectant than from detergent used for hand disinfection. British Journal of Dermatology. 2005;153:1142–1146.

[23] Darbre PD, Charles AK. Environmental oestrogens and breast cancer: evidence for combined involvement of dietary, household and cosmetic xenoestrogens. Anticancer Research. 2010;30:815–827.

[24] Fruijtier-Pölloth C. The safety of synthetic zeolites used in detergents. Archives of Toxicology. 2009;83:23–35.

[25] Johnson GI, Hofman PJ. Postharvest technology and quarantine treatments. In: Litz RE, editor. The Mango. Botany, Production and Uses. 2nd ed. UK: CABI Publ.; 2009. pp. 529–605.

[26] EPA. Conditions for minimum risk pesticides. Environmental Protection Agency (EPA) [Internet]. 2016. Available from: http://www.epa.gov/minimum-risk-pesticides/conditions-minimum-risk-pesticides#tab-2. [Accessed: 2016/01/02].

[27] Curkovic T. Integrated pest management in olive orchards in Chile [in Spanish]. In: Fichet T, Henríquez J, editors. Contributions to the Knowledge to the cultivation of the Olive in Chile. Santiago: Volumen 23 Serie Ciencias Agronómicas Universidad de Chile; 2013. pp. 155–204.

[28] Kovach J, Petzoldt C, Degni J, Tette J. A method to measure the environment impact of pesticides. New York's Food and Life Sciences Bulletin. 1992;139:1–8.

[29] Shiau BJ, Sabatini D, Harwell JH. Solubilization and microemulsification of chlorinated solvents using direct food additive (edible) surfactants. Ground Water. 1994;32:561–569.

[30] Butler GD, Henneberry TJ, Stansly PA, Schuster DJ. Insecticidal effects of selected soaps, oils and detergents on the sweetpotato whitefly. Florida Entomologist. 1993;76:161–167.

[31] Cranshaw WS. Insects control: soaps and detergents. 1996. Available from: http://www.extension.colostate.edu/docs/pubs/insect/05547.pdf [Accessed: 2016/01/02].

[32] Sparks TC, Nauen R. IRAC: mode of action classification and insecticide resistance management. Pesticide Biochemistry and Physiology. 2015;121:122–128.

[33] Simpson SJ, Douglas AE, editors. The insects: structure and function. RF Chapman. 5th ed. USA: Cambridge University Press; 2013. 929 p.

[34] Curkovic T, Gonzalez R, Barría G. Effectivity of a detergent in the control of the black scale *Saissetia oleae* (Oliver) (Homoptera: Coccidae) in grapefruits and oleander [in Spanish]. Investigación Agrícola. 1993;13:43–46.

[35] Santibáñez D. Evaluation of dehydration and epicuticle wax removal as factors associated to mortality on females of Pseudococcus viburni Signoret (Hemiptera:

Pseudococcidae) treated with agricultural detergents [in Spanish]. Santiago: College of Agronomic Sciences, University of Chile; 2010.

[36] Yamazaki K. Gone with the wind: trembling leaves may deter herbivory. Biological Journal of the Linnean Society. 2011;104:738–747.

[37] Curkovic T, Araya JE. Acaricidal action of two detergents against *Panonychus ulmi* (Koch) and *Panonychus citri* (McGregor) (Acarina: Tetranychidae) in the laboratory. Crop Protection. 2004;23:731–733.

[38] Curkovic T, Durán M, Ferrera C. Control of *Brevipalpus chilensis* Baker (Acari: Tenuipalpidae) with agricultural detergents in the laboratory and under field conditions. Chilean Journal of Agricultural and Animal Sciences. 2013;29:73–82.

[39] Saini RK, Yadav PR. Sampling, surveillance, and forecasting of pests. In: Jain PC, Bhargava MC, editors. Entomology Novel Approaches. India: New India Publ. Agency; 2007. pp. 19–42.

[40] Warnock DF, Loughner R. Comparing three solutions for extracting western flower thrips from Coleus shoots. Hortscience. 2002;37:787–789.

[41] Marer PJ, Flint ML, Stimmann MW. The safe and effective use of pesticides. USA: University of California, Statewide IPM Proyect, Division of Agriculture & Natural Resources, Publication 3324; 1988. 387 p.

[42] Ware GW. The Pesticide Book. USA: Thomson Publications; 1994. 366 p.

[43] Szumlas D. Behavioral responses and mortality in German cockroaches (Blattodea: Blatellidae) after exposure to dishwashing liquid. Journal of Economic Entomology. 2002;95:390–398.

[44] Chen J, Shang HW, Jin X. Response of *Solenopsis invicta* (Hymenoptera: Formicidae) to potassium oleate water solution. Insect Science. 2010;17:121–128.

[45] Weinzierl RA. Botanical insecticides, soaps, and oils. In: Rechcigl JE, Rechcigl NA, editors. Biological and Biotechnological control of insect pests. USA: CRC Press; 2000. pp. 101–137.

[46] Rostas M, Blassmann K. Insects had it first: surfactants as a defense against predators. Proceedings of the Royal Society B: Biological Sciences. 2009;276:633–638.

[47] Arias C, Silva G, Tapia M, Hepp R. Asessment of household soaps and detergents to control *Trialeurodes vaporariorum* (Westwood) in the greenhouse [in Spanish]. Agro-Ciencia. 2006;22:7–14.

[48] Asiedu E, Afun JVK, Kwoseh C. Control of *Planococcus citri* (Risso) (Hemiptera: Pseudococcidae) on white yam variety (*Pona*) in storage. Africa Development and Resources Research Institute. 2014;9:1–15.

[49] Ebbing DD. General Chemistry. 4th ed. USA: Houghton Mifflin Company; 1993. 1085 p.

[50] Matsumura F. Toxicology of Insecticides. USA: Plenum Press; 1975. 503 p.

[51] Rust D, Wildes S. Surfactants: A Market Opportunity Study Update. Midland, MI: OmniTech International Ltd.; 2008.

[52] Villar J. Effect of the TS-2035 agricultural detergent used as adjuvant for two entomopathogenic fungi and one conventional insecticide sprayed on adult females of the obscure mealybug (Pseudococcus viburni Signoret) (Hemiptera: Pseudococcidae) in the laboratory [in Spanish]. Santiago: College of Agronomic Sciences, University of Chile; 2015.

[53] Cowles RS, Cowles EA, McDermott AM, Ramoutar D. "Inert" formulation ingredients with activity: toxicity of trisiloxane surfactant solutions to twospotted spider mites (Acari: Tetranychidae). Journal of Economic Entomology. 2000;93:180–188.

[54] Sazo L, Araya JE, de la Cerda J. Effect of a siliconate coadjuvant and insecticides in the control of mealybug of grapevines, *Pseudococcus viburni* (Hemiptera: Pseudococcidae) [in Spanish]. Ciencia e Investigación Agraria. 2008;35:215–222.

[55] Ripa R, Rodriguez F, Larral P, Luck R.F. Evaluation of a detergent based on sodium benzene sulfonate for the control of woolly whitefly *Aleurothrixus floccosus* (Maskell) (Hemiptera: Aleyrodidae) and citrus red mites *Panonychus citri* (McGregor) (Acarina: Tetranychidae) on oranges and mandarins [in Spanish]. Agricultura Técnica 2006;66:115–123.

[56] Valles SM, Capinera JL. Response of larvae of the southern armyworm, *Spodoptera eridania* (Cramer) (Lepidoptera: Noctuidae), to selected botanical insecticides and soap. Journal of Agricultural Entomology. 1993;10:145–153.

[57] Curkovic T, Gonzalez R, Barría G. Control of first instar nymphs of Saissetia oleae (Oliver) (Homoptera: Coccidae) with detergents on grapefruits and oleander [in Spanish]. Simiente. 1995;65:133–135.

[58] Curkovic T, Ballesteros C. Integrated pest management of key pests in olives in Copiapó and Huasco [in Spanish]. In: Fichet T, Razeto B, Curkovic T, editors. The Olive: Agronomic Study in the Atacama región. Santiago: Volumen 16 Serie Ciencias Agronómicas Universidad de Chile; 2011. pp. 117–144.

[59] Hopkins WG. Introduction to Plant Physiology. USA: John Wiley and Sons, Inc.; 1995. 464 p.

[60] Tamura H, Knoche M, Bukovac M. Evidence for surfactant solubilization of plant epicuticular wax. Journal of Agricultural and Food Chemistry. 2001;49:1809–1816.

[61] Morais H, Medri ME, Marur CJ, Caramori PH, de Arruda Ribeiro AM, Gomes JC. Modifications on leaf anatomy of *Coffea arabica* caused by shade of pigeonpea (*Cajanus cajan*). Brazilian Archives of Biology and Technology. 2004;47:863–871.

[62] Ambers L, Cheng S, Steffens L. Surfactant residues in Maryland tobacco treated with a fatty-alcohol-type sucker-control formulation. Animal Reproduction Science. 1983;12:15–19.

[63] Moore WS, Profita JC, Koehler CS. Soaps for home landscape insect control. California Agriculture. 1979;37:13–14.

[64] Krogh KA, Halling-Sorensen B, Mogensen BB, Vejrup KV. Environmental properties and effects of nonionic surfactant adjuvants in pesticides: a review. Chemosphere. 2003;50:871–901.

[65] Osborne LS, Petitt FL. Insecticidal soap and predatory mite, *Phytoseiulus persimilis* (Acari: Phytoseiidae), used in management of the twospotted spider mite (Acari: Tetranychidae) on greenhouse grown foliage plants. Journal of Economic Entomology. 1985;78:687–691.

[66] Sazo L, Astorga I, Araya JE. Effect of sodium disoctil sulfosuccinate on mites *Panonychus ulmi* (Koch) (Tetranychidae) and *Neoseiulus californicus* (McGregor) (Phytoseiidae) on apple orchards in the central zone of Chile. Boletín de Sanidad Vegetal (Plagas) 2005;31:623–630.

[67] Lawson DS, Weires RW. Management of European red mite (Acari: Tetranychidae) and several aphid species on apple with petroleum oils and an insecticidal soap. Journal of Economic Entomology. 1991;84:1550–1557.

[68] Curkovic T, Araya JE, Canales C, Medina A. Evaluation of two agriculture detergents as control alternatives for green peach aphid and twospotted spidermite, two pests affecting peach orchards in Chile. Acta Horticulturae. 2006;713:405–407.

[69] Sclar DC, Gerace D, Tupy A, Wilson K, Spriggs SA, Bishop RJ, Cranshaw WS. Effects of application of various reduced-risk pesticides to tomato, with notes on control of greenhouse whitefly. Hortechnology. 1999;9:185–189.

[70] Puritch GS. The toxic effects of fatty acids and their salts on the balsam woolly aphid, *Adelges piceae* (Ratz.). Canadian Journal of Forest Research. 1975;5:515–522.

[71] Singh SP, Rao NS. Field evaluation of fish oils insecticidal rosin soap against soft green scale, *Coccus viridis* on citrus. Indian Journal of Plant Protection. 1979;7:208–210.

[72] Reimer NJ, Beardsley JW. Epizootic of white halo fungus, *Verticillium lecanii* (Zimmerman), and effectiveness of insecticides on *Coccus viridis* (Green) (Homoptera: Coccidae) on coffee at Kona, Hawaii. Proceedings of the Hawaiian Entomological Society. 1992;31:73–81.

[73] Imenes SDL, Bergmann EC, de Faria AM, Martins WR. Record of high infestation and evaluation of soap solutions on the control of *Protopulvinaria pyriformis* Cockerell, 1894

(Hemiptera: Coccidae) on *Schefflera arboricola* (Hayata) Merr. (Araliaceae) [in Portuguese]. Arquivos do Instituto Biologico Sao Paulo. 2002;69:59–62.

[74] Sabine BN. Insecticidal control of citrus pests in coastal central Queensland. Queensland Journal of Agricultural and Animal Sciences. 1969;26:83–88.

[75] Nielsen DG. Evaluation of biorational pesticides for use in arboriculture. Journal of Arboriculture. 1990;16:82–88.

[76] Hubbard JL, Potter DA. Managing calico scale (Hemiptera: Coccidae) infestations on landscape trees. Arboriculture and Urban Forestry. 2006;32:138–147.

[77] Urías-López MA, Hernández-Fuentes LM, Osuna-García JA, Pérez-Barraza MH, García-Álvarez NC, González-Carrillo JA. Field insecticide sprays for controlling the white mango scale (Hemiptera: Diaspididae) [in Spanish]. Revista Fitotecnia Mexicana. 2013;36:173–180.

[78] Blanco-Metzler H, Zúñiga A. Control of the cycad scale *Aulacaspis yasumatsui* (Hemiptera: Diaspididae) with different commercial brands of soap in Costa Rica [in Spanish]. InterSedes. 2013;14:114–122.

[79] Surulivelu T, Gulsar Banu J, Sonai Rajan T, Dharajothi B, Amutha M. Evaluation of fungal pathogens for the management of mealybugs in Bt cotton. Journal of Biological Control. 2012;26:92–96.

[80] Gill HK, Goyal G, Gillett-Kaufman J. Citrus mealybug (*Planococcus citri* (Rosso) (Insect: Hemiptera: Pseudococcidae). Florida Cooperative Extension Service, Institute of Food and Agricultural Sciences, University of Florida; 2012. Available on: edis.ifas.ufl.edu/in947 (Revised April 2016).

[81] Kumar R, Nitharwal M, Chauhan R, Pal V, Kranthi KR. Evaluation of ecofriendly control methods for management of mealybug, *Phenacoccus solenopsis* Tinsley in cotton. Journal of Entomology. 2012;9:32–40.

[82] Chirinos-Torres L, Geraud-Pouey F, Chirinos DT, Fernández C, Guerrero N, Polanco MJ, Fernández G, Fuenmayor R. Effect of insecticides on *Capulinia* sp. near *jaboticabae* von Ihering (Hemiptera: Eriococcidae) and its natural enemies in Mara county, Zulia state, Venezuela [in Spanish]. Boletín de Entomología Venezolana. 2000;15:1–16.

[83] Cisterna V, Sepulveda G. Efficiency of an enzymatic detergent in the control of *Praelongorthezia olivicola* (Beingolea) (Hemiptera: Ortheziidae) in the northern Chile. Revista de la Facultad de Ciencias Agrarias de la Universidad Nacional de Cuyo. 2015;47:213–218.

[84] Contreras J, Quinto V, Abellán J, Fernández E, Grávalos C, Moros L, Bielza P. Effects of natural insecticides on *Frankliniella occidentalis* and *Orius* spp. IOBC/wprs Bulletin 2006;29:331–336.

[85] Karunamoorthi K, Bishaw D, Mulat T. Laboratory evaluation of Ethiopian local plant *Phytolacca dodecandra* extract for its toxicity effectiveness against aquatic macroinvertebrates. European Review for Medical and Pharmacological Sciences. 2008;12:381–386.

[86] Sholberg P, Boule J. Palmolive detergent controls apple, cherry, and grape powdery mildew. Canadian Journal of Plant Sciences. 2009;89:1139–1147.

[87] Stewart CD, Braman SK, Sparks BL, Williams-Woodward JL, Wade GL, Latimer JG. Comparing an IPM pilot program to a traditional cover spray program in commercial landscapes. Journal of Economic Entomology. 2002;95:789–796.

[88] Lee CY, Lo KC, Yao MC. Effects of Household Soap Solutions on the Mortality of the Two-spotted Spider Mite, Tetranychus urticae Koch (Acari: Tetranychidae). Formosan Entomology. 2006;26:379–390.

[89] Raffa KF, Priester TM. Synergists as research tools and control agents in agriculture. Journal of Agricultural Entomology. 1985;2:27–45.

[90] Haviland DR, Bentley WJ, Daane KM. Hot-water treatments for control of *Planococcus ficus* (Homoptera: Pseudococcidae) on dormant grape cuttings. Journal of Economic Entomology. 2005;98:1109–1115.

[91] Follett P, Neven L. Current trends in quarantine entomology. Annual Review of Entomology. 2006;51:359–385.

[92] Carpio C. Basis for integrated pest management of the obscure mealybug (Pseudococcus viburni) in pomegranates: evaluation of monitoring and control methods [in Spanish]. Santiago: College of Agronomic Sciences, University of Chile; 2013.